LINDESNES LIGHTHOUSE

LANDEGODE LIGHTHOUSE

FÆRDER LIGHTHOUSE

ÅSVÆR LIGHTHOUSE

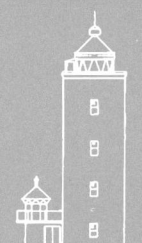
FRUHOLMEN LIGHTHOUSE

GULLHOLMEN LIGHTHOUSE

SLETTNES LIGHTHOUSE

JOMFRULAND LIGHTHOUSE

BASTØY LIGHTHOUSE

FUGLØYKALVEN LIGHTHOUSE

The Norwegian Coastal Current
– Oceanography and Climate

The Norwegian Coastal Current
– Oceanography and Climate

Editor:
Roald Sætre

Illustrations:
Karen Gjertsen and Kareen Bröker

Graphic design and technical editing:
Harald E. Tørresen

INSTITUTE OF MARINE RESEARCH
HAVFORSKNINGSINSTITUTTET

tapir academic press

© Tapir Academic Press, Trondheim 2007

ISBN: 978-82-519-2184-8

This publication may not be reproduced, stored in a retrieval system or transmitted in any form or by any means; electronic, electrostatic, magnetic tape, mechanical, photo-copying, recording or otherwise, without permission.

Graphic design and technical editing:
Harald E. Tørresen, Institute of Marine Research

Printed by:
PDC Tangen

Photo cover:
From Golten/Sotra, Western Norway
Photo: Per Bækken

Tapir Academic Press
N–7005 TRONDHEIM
Tel.:   + 47 73 59 32 10
Fax: + 47 73 59 32 04
E-mail: forlag@imr.no
www.tapirforlag.no

Contents

  Preface 7

1. **Introduction** 9
   Roald Sætre

2. **Studies of the coastal region** 19
   Background and history
   Roald Sætre

3. **The origin of the coastal zone** 35
   Terje Thorsnes and Oddvar Longva

4. **Driving forces** 45
   Roald Sætre

5. **Properties of the coastal water masses** 59
   Roald Sætre

6. **Temporal and spatial distribution of nutrients** 73
   Francisco Rey, Jan Aure and Didrik S. Danielssen

7. **Short-term variability and small-scale features** 89
   Roald Sætre

8. **Characteristic circulation features** 99
   Roald Sætre and Jan Aure

9. **Coast/fjord water exchange** 115
   Jan Aure, Lars Asplin and Roald Sætre

10. **Climate changes in the Norwegian Coastal Current** 125
    Roald Sætre, Jan Aure and Didrik S. Danielssen

11. **Operational oceanography** 139
    – Challenges and possibilities
    Johnny A. Johannessen, Bruce Hackett, Einar Svendsen, Henrik Søiland,
    Lars P. Røed, Nina Winther, Jon Albretsen, Didrik S. Danielssen,
    Lasse Petterson, Morten Skogen and Laurent Bertino

12. **References** 151

  The Authors 159

Preface

At the end of the project "The Norwegian Coastal Current" in 1980, it was proposed to produce a popular scientific book on the physical/chemical conditions of the Norwegian Coastal Current. In spite of enthusiasm on the part of the potential authors, this work never got off the ground.

The research programme on polar ecology *"Pro Mare"* was carried out during 1984–1989. Efforts concentrated on the Barents Sea, with the main objective of increasing our knowledge of how the marine Arctic ecosystem functioned. Based mainly on the results from this programme, a book on the Barents Sea Ecosystem was published in Norwegian in 1992.

In 1994 the Norwegian research programme on North Norwegian Coastal Ecology (MARE NOR) came to an end, and as a part of the reporting procedure, a small popular scientific book in Norwegian describing the main features of the coastal ecosystem, including the physical environment, was published.

The Institute of Marine Research carried out a research programme *"Mare cognitum"* on the Norwegian Sea ecosystem in 1993–2001. Scientists from several countries participated in the programme. Drawing on the results from *"Mare cognitum"* "The Norwegian Sea Ecosystem" was published in 2004, as a summary of the current knowledge of that ecosystem. As a follow-up of that work, it was proposed to produce a similar book for the Norwegian coastal ecosystem. However, due to competition from other activities, it was difficult for several key persons to find the time to participate in this work. As a compromise, and as a starting point for a future summary of the ecology of Norwegian coastal waters, it was decided to limit the book to the physical/chemical conditions of the Coastal Current. Actually, interannual fluctuations in the physical conditions are most likely to be the prime cause of ecosystem variability.

The aim of this book is to summarise our current knowledge of the physical/chemical properties and dynamics of the Norwegian Coastal Current in such a way as to make the material more easily understandable for non-experts. As editor, I thank the contributing authors. Special thanks to Karen Gjertsen, Kareen Bröker, Øyvind Strand and Jaime Alvarez for their skilled work with the illustrations, Ingunn Bakketeig for quality assurance and corrections and Hugh Allen for help in improving the language. Harald E. Tørresen has been responsible for the technical editing and layout of the book.

Bergen, November 2006

Roald Sætre
Editor

Introduction

Roald Sætre

1.1 General description

The Norwegian Coastal Current originates primarily from the fresh water outflow from the Baltic and the fresh water runoff from Norway. This water mixes with the North Sea Water and Atlantic Water and flows northwards along the coast of Norway as a wedge-shaped low-salinity current bordered by the Norwegian North Atlantic Current off the central and northern coasts of Norway (Figure 1.1). A branch of the Atlantic Water enters the North Sea and penetrates the Skagerrak, where it turns back and leaves the North Sea below the Norwegian Coastal Current. As it mixes with the Atlantic Water the salinity of the Norwegian Coastal Current gradually increases and its stratification decreases along its route. For the most part, the Norwegian Coastal Current is driven by its density structure, which is mainly determined by the salinity distribution. However, there is often a significant barotropic component or current component caused by changes in the surface elevation, due to piling up of water along the Norwegian coast during persistent southwesterly winds. Typical current speeds are 20-50 cm/s, with maximum speeds exceeding 100 cm/s and typical transport

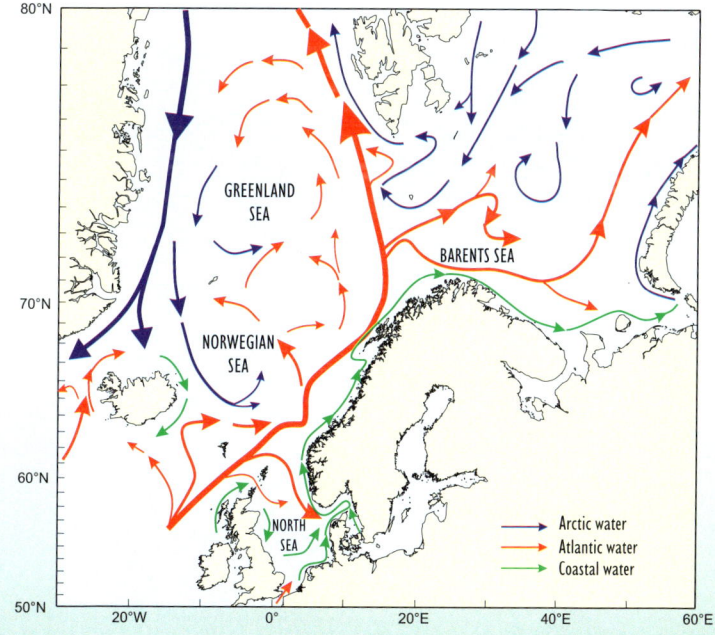

▶ Figure 1.1
The general circulation of the North Sea, the Nordic Seas and the Barents Sea.

▶ Figure 1.2
The Torungen lighthouse on the Skagerrak coast during a "blond" summer night.

values of 1-2 Sverdrup (10^6 m³/s). The range of variation is approximately of the same order as the mean value. The Norwegian Coastal Current transports the freshwater outflow from the North Sea, the Baltic and the Norwegian coast into the Barents Sea and is thus one of the major sources of freshwater in the Arctic (Figure 1.1).

Norvegr or Norway was originally the name of the long northern coast. The shortest distance between the southernmost point on the Norwegian coast to the northernmost equals that from the southernmost to Rome. The length of the coastal line measured along the mainland is about 24,000 km or more than half the distance around the Equator, while the coastlines of the numerous islands come to about 60,000 km. A barrier of islands, islets and skerries protects the area of sounds, bays and fjords within it from exposure to the forces of the open sea. Along most of the coast the mainland does not actually reach the edge of the ocean. Deep fjords cut into the mountainous inner part of the country. The longest of these, the Sognefjord, penetrates 200 km inland from the coastline. The Coastal Base Line (CBL) is a line drawn between defined points on the outermost islands or skerries along the coast. Following the EU Water Framework Directive the coastal zone is defined as the area lying within a line one nautical mile outside the CBL. The coastal zone thus covers an area close to 100,000 km², which is equivalent to nearly one third of the mainland.

1.2 Function and activities

Eighty percent of the Norwegian population lives within 10 km of the coast. There are two main reasons for this – availability of food and communications. The coastal zone has long offered a stable food supply from both stationary and migratory fish stocks. It also facilitated communications by safe sailing in protected areas both in and out of the fjords as well as north–south along the coast, while the mountainous inner part of the country hindered transport and communications. For this reason, the coastal waters have always been regarded as Highway no 1 in Norway.

▶ **Figure 1.3** Spawning location and seasonal migrations of some of the commercially most important fish stocks.

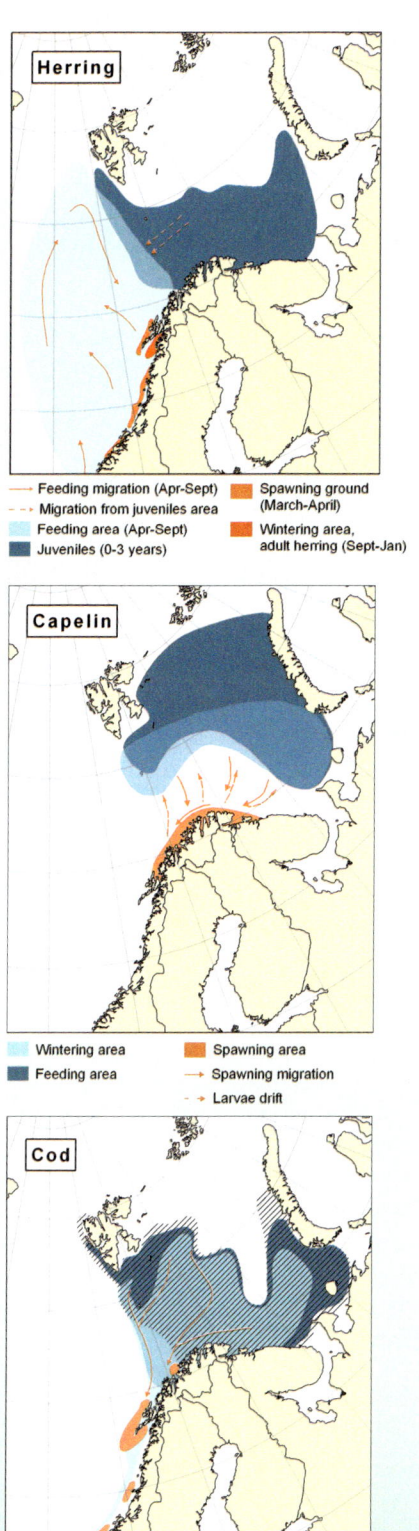

Norway is among the most important fishing nations in the world. During the past ten years annual Norwegian catches of fish have totalled around three million tonnes. The Norwegian coastal area is the spawning ground for a number of important oceanic fish stocks such as cod, herring and capelin, giving their young good chances of survival. The eggs and larvae drift northwards with the coastal current towards the nursery and feeding grounds in the Barents Sea. Huge numbers of fish on spawning or feeding migrations reach the Norwegian coast every winter and spring and become the objects of seasonal fisheries. This migration of fish means that there is an enormous transport of biomass into the coastal ecosystems (Figure 1.3). The spawning products of these oceanic stocks alone are equivalent to 10 to 20 times the weight of the total Norwegian human population. The seasonal fisheries, especially for cod and herring, formed the backbone of the coastal economy, and fisheries-related activities were the foundation of most coastal communities. The results of the fisheries could vary due to natural variations in year-class strength. People on the Norwegian coast were quite familiar with great inter-annual fluctuations in the fisheries. They were particularly concerned about the variations in the herring fishery or the so-called "herring periods", and in the course of the centuries various explanations were offered. In earlier times these were associated with the quality of the king; a good king meant a rich harvest from the sea and the land, while a bad king had the opposite effect. Later, many people believed that failure of the fisheries was a punishment from God for their sinful lives.

With its huge areas of sheltered sea and its mild oceanic climate due to the influence of the Atlantic water, the Norwegian coast is ideal for farming of marine organisms. The early

▶ Figure 1.4
Kayaking in the Nærøyfjord.

Photo: Øyvind Heen

1970s saw a breakthrough for Atlantic salmon farming in Norway, since when there has been a steep increase in production until in 2000 the production of farmed salmon reached 470,000 tonnes. In the past few years production has remained approximately constant, i.e. around 0.5 million tonnes. Farming of other species, such as cod and halibut, is still in its infancy and will probably develop significantly in the next few years. The same goes for crustacean and shellfish farming.

The offshore oil and gas industry has become a major economic activity off the Norwegian coast since the mid-1960s. The start of production on the Ekofisk field in the North Sea in 1971 was a major event in Norwegian economic history. In 1980 the continental shelf north of 62°N was opened for exploration of its oil and gas resources, and today the areas off the central and north Norwegian coast are the most interesting ones for new development projects. While the last century was the century of oil, we have now passed into the century of gas, when Norwegian gas production will exceed that of oil.

In spite of the high level of activity in the coastal zone a substantial part of it can probably still be regarded as Europe's last marine wilderness. The Norwegian coast is therefore also important for recreation, both for the local people and for tourists. In the late 19th century, rich foreign tourists, especially from the UK and Germany, started to spend their vacations in Norway, although their numbers remained low until the growing wealth of the western world after World War II encouraged the development of the tourist industry. All-season recreational fishing has become very popular among foreign tourists. Around 200,000 people, mostly Germans, come to Norway every year in order to fish. The total catch of recreational fisheries has been estimated at around 15,000 tonnes. Cruises along the coast by the coastal express liners are also very popular with tourists. Every day throughout the year a ship leaves Bergen for a trip to Kirkenes close to the Russian border and back to Bergen. This round trip takes 11 days. In the course of the past few years these vessels have been upgraded to a standard close to international cruise vessel level. I 2004, part of the outer coast, the Vega archipelago off northern Norway (Figure 1.6), was put on UNESCO's World Heritage List and in 2005 two fjords in western Norway, the Nærøyfjord (Figure 1.4) and the Geirangerfjord (Figure 1.5), were added to this list

1.3 Research and management

There are many interests involved in using or even exploiting the coastal zone and these interests may well come into conflict. Some human activities may have adverse effects on the marine ecosystem. In order to balance the needs for use and conservation of the coastal zone, the Coastal Zone Planning and Management process was introduced. Fisheries management has seen a movement from the management of stocks of single species to management at ecosystem level. Both these processes demand in-depth knowledge of the physical conditions of the area involved, i.e. the chemical/physical oceanography of the coastal zone.

Norway is currently in the process of implementing the EU Water Framework Directive (WFD). The main objective of this directive is to protect all water resources, including coastal waters out to one nautical mile from the Coastal Base Line. The aim is that all water bodies should have "good ecological status" by the end of 2015. Measures may be taken to deal with any body of water that does not achieve such status. Norwegian coastal waters are

▶ **Figure 1.5**
View over the Geirangerfjord an early morning in May with fruit trees along the fjord surrounded by snow-capped mountains.

Photo: Finn Magne Grande

divided into a number of different "ecological types". Four ecological regions, the Skagerrak, the North Sea, the Norwegian Sea and the Barents Sea, have been established (Figure 1.7), and within each of these regions, individual water bodies are identified according to different nature categories, such as degree of exposure, fresh-water influence, stratification and current conditions. This work is in progress, and the interim results indicate that the total number of water bodies along the Norwegian coast will probably approach 2000. The WFD also requires the various water types to be monitored, in order to enable the authorities to document their condition and any observable trends. Designing a relevant and cost-efficient monitoring programme demands a thorough knowledge of both the properties and the dynamic structure of the coastal waters.

In recent years there has been a trend towards taking a more holistic view of the management of the marine environment and its resources. The term "an ecosystem or ecological approach to management" has been widely used. The Biodiversity Convention of 1992 defines an ecosystem as "a dynamic complex of plant, animal and micro-organism communities and their non-living environment, interacting as a functional unit". In simpler terms, an ecosystem comprises both the non-living environment and all living organisms within a defined geographical area. The organisms interact with the physical environment and with each other. The physical environment is an important part of marine ecosystems. Bottom depth and bottom type, temperature, salinity and currents all determine both which organisms can survive in the ecosystem and where they can live.

▶ **Figure 1.6**
Aerial view of the Vega Archipelago.

Photo: Torild Wika

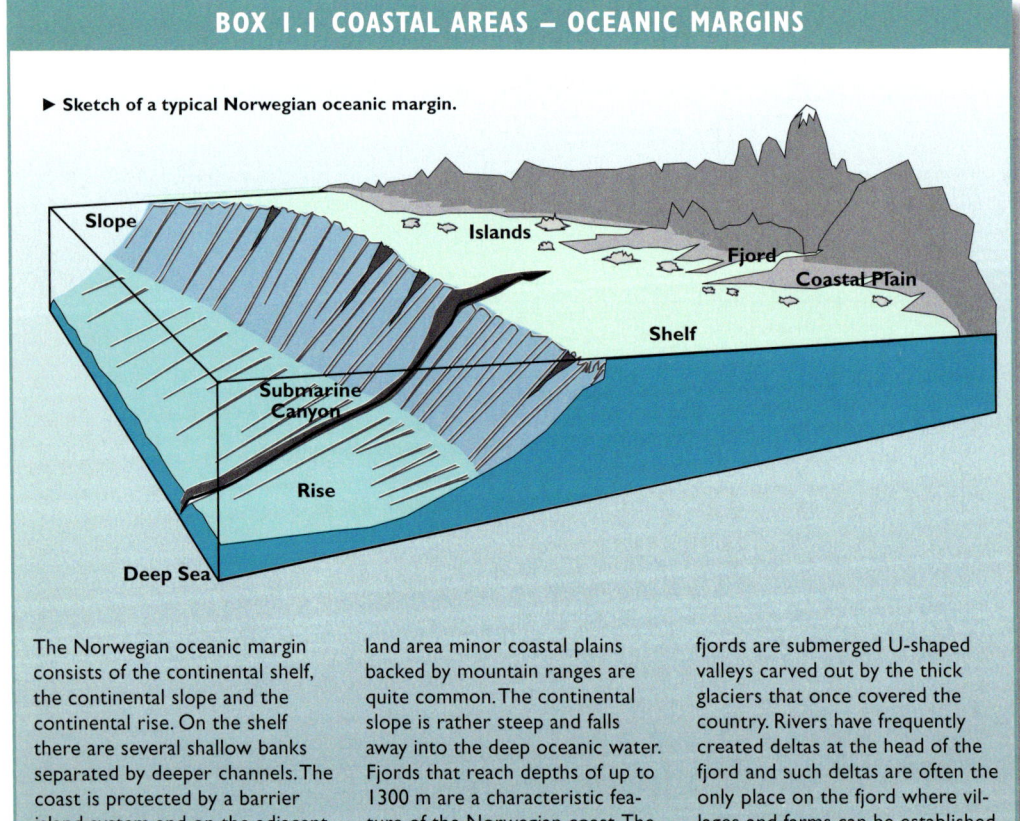

BOX 1.1 COASTAL AREAS – OCEANIC MARGINS

▶ Sketch of a typical Norwegian oceanic margin.

The Norwegian oceanic margin consists of the continental shelf, the continental slope and the continental rise. On the shelf there are several shallow banks separated by deeper channels. The coast is protected by a barrier island system and on the adjacent land area minor coastal plains backed by mountain ranges are quite common. The continental slope is rather steep and falls away into the deep oceanic water. Fjords that reach depths of up to 1300 m are a characteristic feature of the Norwegian coast. The fjords are submerged U-shaped valleys carved out by the thick glaciers that once covered the country. Rivers have frequently created deltas at the head of the fjord and such deltas are often the only place on the fjord where villages and farms can be established.

In the ocean areas around Norway three so-called Large Marine Ecosystems (LMEs) have been defined (Sherman and Skjoldal, 2002); the Norwegian Sea, the Barents Sea and the North Sea (Figure 1.1). The criteria for this definition are based on bottom topography, water mass properties and current conditions, in addition to biological production. LMEs should be really large, typically 200,000 km² or larger. The spatial scales of the large commercially fish stocks are an important reason why LMEs are usually relatively large in scale. The three LMEs surrounding Norway are open systems; the ocean currents flow through them and transport plankton organisms in and out of them. Organisms that inhabit any one of the systems have free access to the neighbouring system. However, a number of fish populations live their whole life within a system while there are also important populations that spend parts of their life cycle outside their "own" area.

The coastal zone can be regarded as forming a part of each of the three large ecosystems, but also as consisting of a number of smaller ecosystems. A fjord could be regarded as an ecosystem because most of its organisms and the most important processes that take place within it interact as a functional unit. However, all the possible coastal ecosystems depend on and interact with the large oceanic ecosystems. In addition to the large oceanic fish populations, which have their spawning and nursery grounds on the coast, there are also a number of local populations of fish and other marine organisms that spend their whole life on the coast. There is no natural boundary for the coastal zone. As we have seen, the EU Water Framework Directive defines it as one nautical mile outside the Coastal Base Line. In this book we treat the whole area covered by the Norwegian Coastal Current as coastal zone. That means that the whole of

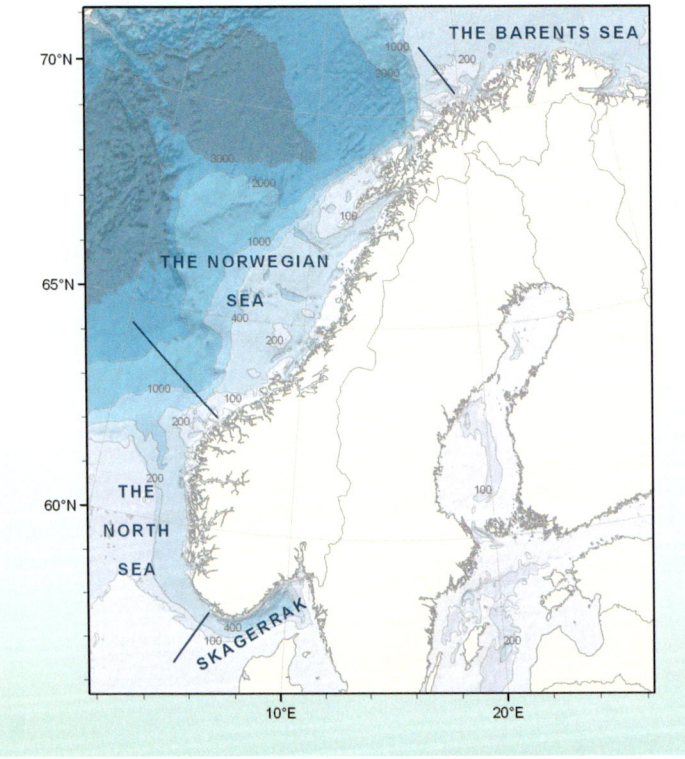

▶ Figure 1.7
The four ecological regions in Norway as defined for the purpose of the EU Water Framework Directive.

BOX 1.2 MARINE PROTECTED AREAS IN NORWAY

A governmental advisory committee recently made recommendations regarding the establishment of Marine Protected Areas in Norway. They proposed the inclusion of 36 areas in the first phase of a Norwegian national marine conservation plan. The areas concerned ranged in size from 5 to 3450 km², and most of them lie within the coastal zone. Representativeness and distinctive qualities were the most important selection criteria for these areas, which will also serve as reference areas for future research and monitoring processes. The figure shows the distribution of the proposed areas for southern Norway. The range of diversity of species in the Norwegian coastal zone is only superficially known. The conservation value of most of the proposed areas lies in the plant and animal life of the seabed, for which reason, the form and character of the seabed have guided the committee's selection of sites. The purpose of conservation is to maintain the subsea landscape or seascape and the species that live there. However, a reasonable balance between conservation and use should be the aim and sustainable use of living resources using methods that do not harm the landscape should therefore be permitted.

Norwegian marine nature may change considerably in the future due to global climate change, new introduced species, new pollutants and greater pressure on the coastal zone due to increased human activity. It will be a demanding task to distinguish between natural and human causes of changes in the future. For that purpose the proposed Marine Protected Areas will also cover the need for reference areas for long-term monitoring and research.

▶ Geographical distribution of the areas in southern Norway proposed for inclusion in the marine conservation plan.

the Norwegian continental shelf and continental slope towards the great ocean depths are included.

In 2004, the Institute of Marine Research (IMR) in Bergen was reorganised in order to meet the challenge of an ecological approach to management of the coastal and oceanic areas within the Norwegian Exclusive Economic Zone (EEZ). The institute has set up Management Advice Programmes (MAP) for each of the three large oceanic ecosystems; the Barents Sea, the Norwegian Sea and the North Sea, as well as for the coastal zone. The MAP for the coastal zone will focus on four areas of research and development. In addition to the EU Water Framework Directive these are:

- *Research on ecosystem-based management*

Around 95 % of living marine organisms are coastal, and there is a need to upgrade our knowledge of the stock structure, dynamics, ecological conditions and total harvesting of the coastal resources.

- *Coastal Atlas*

It is intended to publish a "Coastal Atlas" that will include information about upwelling, productivity indices, local living marine stocks, spawning grounds, etc.

- *Conservation of biological diversity*

Strategic reference areas will be established with a view to arriving at an integrated understanding of the ecological processes involved in the system. In order to secure marine habitats and biodiversity, new fish-capture gear that takes such processes into account will have to be developed.

The basis for the activities in all these areas will be updated knowledge of the characteristics and dynamics of the coastal waters and their temporal and spatial variability.

The target readership for this book is broad; marine scientists – both national and international, students, marine industry and fishery, management personnel and decision-makers and, we hope, also interested members of the general public. The book reviews current knowledge of physical/chemical conditions in the Norwegian Coastal Current. A detailed description of the physical conditions of fjords is regarded as being outwith the scope of this book. However, the general exchange processes between the outer coast and the fjords will be included. The book is intended to be a hybrid of popular description and purely scientific presentation, and we trust that its authors will succeed in this difficult task.

Studies of the coastal region
Background and history

Roald Sætre

2

2.1 The period before the Second World War

The first description of physical conditions on the Norwegian coast is probably to be found in Kongespeilet (The Royal Mirror), which was written around 1250, most likely by an archbishop in Trondheim. The book is a textbook on trading and courtly behaviour in the upper social classes, but it also contains information about seasonal variations in wind patterns, tides and tidal currents, sailing advice for mariners as well as information on the spawning behaviour of fish.

The Swedish archbishop Olaus Magnus Gothus of Uppsala published the first map of the Nordic Seas in 1539 (Carta Marina), which included descriptions of some fantastic sea creatures (Figure 2.1). Later, Peder Claussøn Friis (1545-1614), Petter Dass (1647-1707) and Erich Pontoppidan (1698-1764) all provided descriptions of various aspects of physical conditions on the Norwegian coast, based mostly on second- or third-hand sources. In his "Natural History of Norway" of 1755, the bishop of Bergen, Erich Pontoppidan, describes the bathymetry along the coast: *"The bottom of the sea is here, as everywhere, full of inequalities, and in this respect, not less varied than the land, which is frequently an alternate succession of high mountains, and deep valleys. The analogy is the same in the substance of the bottom of the sea"*. Pontoppidan also deals with the salinity of the sea and specifically with the regional differences in salinity be-

▶ **Figure 2.1** Detail from the Olaus Magnus' "Carta Marina" from 1539 off northern Norway.

tween the Baltic, the Norwegian coast and off Iceland. He describes the tidal currents and paid a great deal of attention to the Maelstrom or the Moskenes Current in the Lofoten area (Gjevik et al., 1997; Sætre, 2004).

The depth of the ocean is important both for fisheries and for navigation. In coastal waters, with occasionally strong currents, taking soundings could be a difficult task from a sailing vessel. In 1866 the Norwegian Geographical Survey acquired the combined sail and steam vessel "Hansteen". For the following thirty years, this vessel was used to survey the seabed off Norway and also for fishery research studies (Sars, 1879).

In 1859 Norway initiated practical scientific studies of fisheries, with the marine zoologist Georg Ossian Sars being responsible for the cod fisheries. The investigations focused on how natural conditions such as temperature and currents influenced the behaviour and migration of fish (Sars, 1879, Sundby, 1980). These fishery studies thus provided the justification for the physical oceanographical investigations of Norwegian coastal waters. Ever since those days, close cooperation between physical oceanographers and marine biologists has been a Norwegian tradition. Sars began temperature observations in the Lofoten area, the main spawning field for the cod, in 1878 (Figure 1.3). In 1891 and 1892 the whole Lofoten area was covered by systematic temperature observations, which concluded that the spawning cod tended to prefer temperatures between 4° and 6 °C (Gade, 1894).

Sars (1879) had quite correct ideas about the influence of the Atlantic water and the Gulf Stream in the Lofoten area, as proof of which he mentioned the frequent landings of drifting tropical nuts in the area. However, his concept of a southwesterly flowing Polar current that transported Arctic waters into the Lofoten area was a misunderstanding. This was clearly demonstrated by the results obtained by the Norwegian North-Atlantic Expedition of 1876-1878 (Mohn, 1887), which found a northerly current along the whole Norwegian coast. Mohn presented the first current map of the Nordic Seas calculated by the mean wind situation as well as the internal distribution of water masses. The map is partly biased by Mohn's imprecise observation tools. However, the northward flowing current along the coast of Norway, including the Norwegian Coastal Current, is conspicuous (Figure 2.2). The justification and motives for the rather expensive Norwegian North-Atlantic Expedition seem to have been a combination of purely scientific motives, practical-economic motives related to fisheries, navigation and weather forecasts as well as national political motives (Bjørnsen, 2003).

Mohn (1887) summarised the distribution of temperatures on the Norwegian coastal banks and in the fjords on the basis of all the available measurements, both Norwegian and foreign, including the observations from the survey vessel

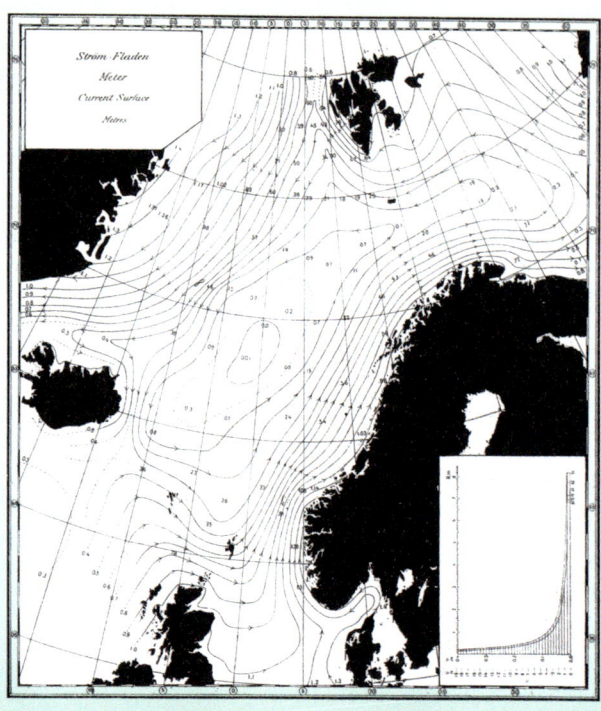

▶ **Figure 2.2**
Current map based on data from the Norwegian North-Atlantic Expedition 1876-1878 (Mohn, 1887).

"Hansteen". He analysed the spatial and seasonal variations in this material and suggested some explanations for the variability in the surface layer. For the conditions in the deeper layers, however, he concluded: *"The material before me will not suffice to explain the proximate cause of the variations in the temperature on the banks and in the depths of the fjords shown by the observations to have occurred there. Meanwhile, the positive result proves the existence of variations apparently unperiodical in nature."* Mohn mentioned that the temperature of the surface layer could propagate downwards by diffusion or by vertical convection if a saline surface layer is cooled and sinks, and he suggested quite correctly that the deeper part of the Skagerrak might be the site of such a process.

The first oceanographic studies of the Skagerrak started in 1869 (Ekman, 1870). A comprehensive survey of all the available data by Pettersson and Ekmann (1891) outlined the basic elements of the water mass distribution and circulation features of the Skagerrak. In a later work the authors further elaborated these elements (Pettersson and Ekman, 1897).

Hjort and Gran (1899) presented a good overview of the physical oceanography of the Norwegian coastal area. They dealt with both seasonal and interannual hydrographic variations in the coastal region (Figure 2.3). Following the suggestion of Pettersson and Ekman (1891) they used the 35 isohaline as a boundary between coastal water and Atlantic waters (Figure 2.4). Hjort and Gran (1899) discussed the hypothesis advanced by Pettersson and Ekman (1891), that water of salinity between 34 and 35 could be of Arctic origin and transported by the East-Icelandic Current to the Norwegian coast around 62–63°N as well as into the North Sea. The hypothesis was mainly based on biological conditions such as the occurrence and distributions of more arctic plankton species. Hjort and Gran (1899) summarised the discussion in the following words:

"We have come to the conclusion that the changes in the Norwegian coastal water are dependent, mainly, on the Gulf Stream on the one hand, and, on the other, on local causes. The Polar current has no direct influence; but our inves-

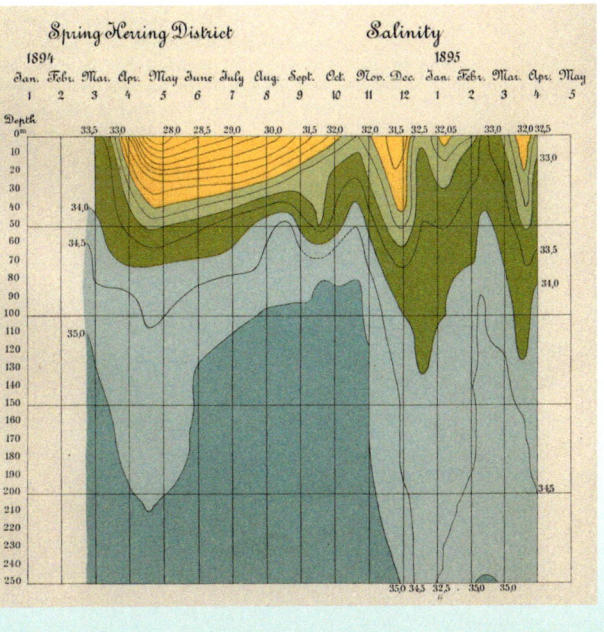

▶ **Figure 2.3**
Isoplet diagrams of salinity from February 1894 to May 1895 off Western Norway (Hjort and Gran, 1899).

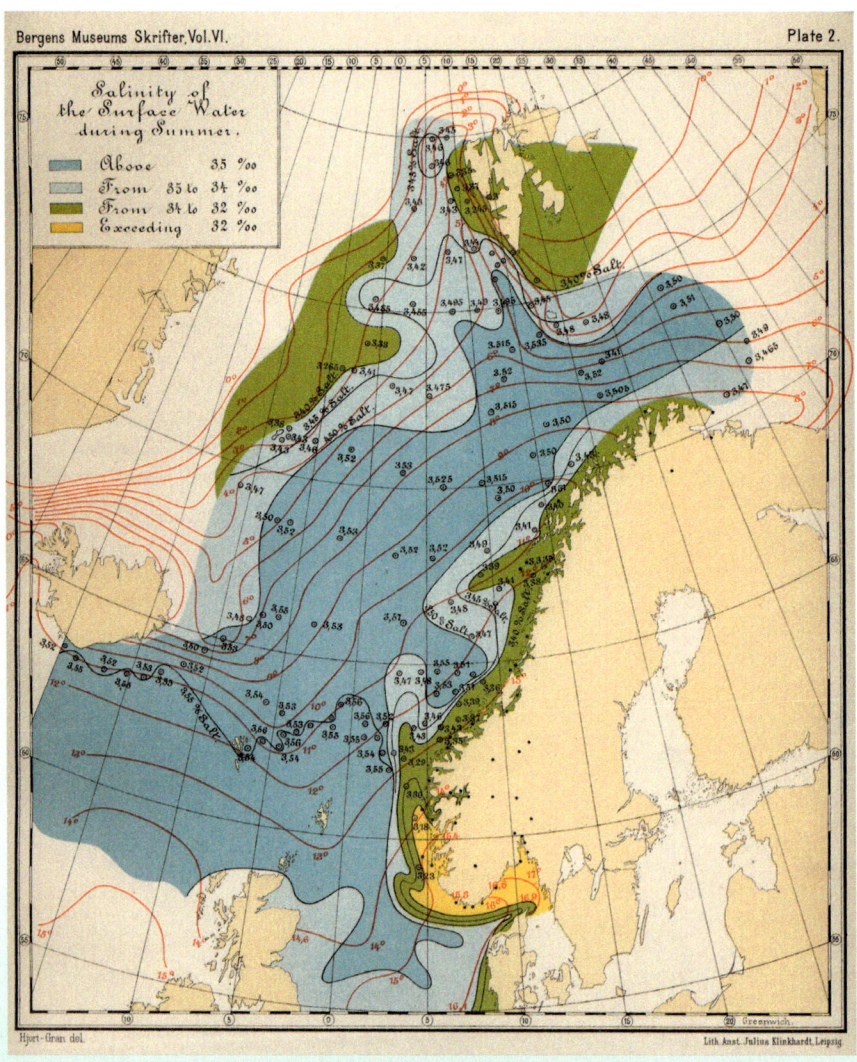

▶ Figure 2.4
Composite temperature and salinity distribution based on observations from the North-Atlantic Expedition (Mohn, 1887) and the German "Drache" Expedition (Anon, 1886) published by Hjort and Gran (1899).

tigations prove that it is of a thickness, which, indirectly, must have the greatest influence on the entire current-system of the North Atlantic and the climate of Europe. And its importance is still further increased through the periodical variations in its thickness and distribution".

In spite of this statement, however, it seems as Gran is not fully convinced, because in a paper published the following year, Gran (1900) writes: *"...there are several circumstances which indicate that the water of the east Icelandic polar current at the coldest time of the year, may be in direct communication with the Norwegian coast-water off Stad and that probably the water of the polar current unites all the year through with that portion of the Gulf Stream which moves from the Faro and Shetland channel north-east to northern Norway and Spitsbergen".*

Hjort and Gran (1899; 1900) dealt with seasonal variations in the coast water, e.g. that the coastal water penetrated to a greater depth during the winter than the summer. Helland-Hansen and Nansen (1909) demonstrated the seasonal lateral displacement of the coastal water: The coastal water wedge was deep and narrow during the winter and wide and thin in the summer. Figure 2.5, which shows the mean seasonal surface salinity variation along the Bergen–Newcastle shipping route, is a good example of the seasonal lateral displacement of the coastal water.

Helland-Hansen and Nansen (1909) were the first to provide a more complete view of what they called "The Norwegian Coast Water" or "The Norwegian Coast Current". According to these authors: *"The coast water is on the whole moving along the coast of Norway, as a continuation of the Baltic Current, from the Skagerrak to the Barents Sea ... As the salinity of the upper layers increases northwards and that of the deeper layer decreases, so as to produce a vertical distribution of salinity which is much more uniform in the north than in the south ..."*

▶ Figure 2.5
Mean surface salinity variations along the Bergen–Newcastle shipping route.

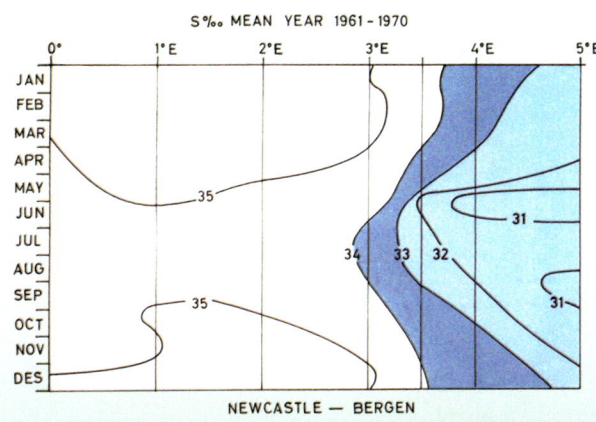

Their current map (Figure 2.6) offered a relatively detailed picture of the circulation in the Norwegian coastal area. Helland-Hansen and Nansen also demonstrated that the outcome of the fishery for sprat and juvenile herring in the Norwegian coastal areas largely depended on the physical conditions of the Norwegian Coastal Current.

In the years between the two world wars, Norwegian marine science experienced a severe reduction in funding. In spite of this, physical oceanography blossomed at the newly established Geophysical Institute headed by Bjørn Helland-Hansen (Mork, 1989). However, most of these activities were not related to the Norwegian coastal areas or the Norwegian Sea. Between 1909 and 1931 physical oceanography research related to fisheries was practically non-existent except for the work of Oscar Sund, which will be touched upon later. One reason for this situation may have been the poor personal relationships between Helland-Hansen and Nansen on one side and Johan Hjort on the other. Johan Hjort resigned from his position as director

▶ Figure 2.6
Current map based mainly on data from RV "Michael Sars" 1900–1904. (Helland-Hansen and Nansen, 1909).

BOX 2.1 CURRENTS AND DRIFTING OBJECTS

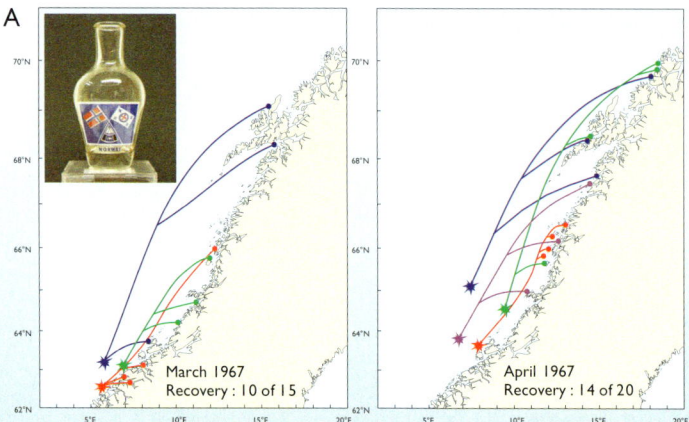

▶ Drift map for IMR drift bottles released in March and April 1967. Five bottles were deployed in each position. The line is not trajectories but only connect the positions of deployment and recovery. The photo shows a drift bottle used by IMR until around 1970.

One hundred years ago Norwegian scientists started to map the currents along the Norwegian coast by deploying drift bottles. Later, other types of drifting objects were used, such as plastic envelopes for simulating oil drift and umbrella-like drifters to map seabed currents. These observations gave only the positions of deployment and recovery, but not trajectories and the drifting speeds of the objects. Figure A shows some results from the use of drifting bottles from 1967.

On the coast, especially north of 62°N, timber, seeds, nuts and parts of plants of more southern and tropical origin are regularly observed. The plant parts most frequently observed on the Norwegian coast are the "sea beans", which are nuts from the Caribbean. These observations of drifting timber and plant parts demonstrate a direct connection to the Norwegian coast from the northeast coast of the USA and Canada as well as to warmer areas, such as the Gulf of Mexico, the Caribbean and Northwest Africa.

Another drifting object often found on the Norwegian coast is the pumice stone. This is a foam-like, light volcanic glass that is capable of floating on water for months. The origin of most of the floating pumice stone observed north of 62°N is probably volcanic activity on Iceland. Occasionally, golf balls from Scotland could be found along the western Norwegian coast (Helland-Hansen, 1930). Modern golf balls sink, so these must have been of a lighter type.

During the First World War hundreds of German and British mines drifted to the Norwegian coast. Most of these had been deployed off the British and the German North Sea coast, particularly during the first year of the war in autumn 1914. These observations of mines along the Norwegian coast demonstrated nicely how the Norwegian Coastal Current transports drifting objects at a rate of 10 to 20 kilometres a day (Nordgaard, 1915).

Eggvin (1940) reported on the drift of a less traditional drifting object. 25 September 1939 a person fell overboard from a coastal vessel off Risør on the Skagerrak coast. Twenty-seven days later the body was found drifting north of Bergen. Eggvin calculated the average drifting speed of the body to be 22.4 cm/s, which later observations confirm to be close to the typical speed of transportation of the coastal current for this part of the coast. During the Second World War another drifting body provided more information about the speed of the Norwegian Coastal Current. During an air attack on Bergen 12 January 1945 a British Lancaster bomber was shot down. On 12 March one of the crew was found drifting in the sea off Nesna on the central Norwegian coast (Figure B). This suggests a drifting speed of 11 cm/s or 9.5 kilometres/day, which is approximately the expected speed of transportation along this part of the coast.

▶ Red: Drift of a drowned person's body from Risør to Hernar north of Bergen. Blue: Drift of Royal Air Force pilot's body from Marsteinen off Bergen to Nesna.

of fisheries in 1917, thus depriving marine science of some of its driving force vis-à-vis the funding authorities.

2.2 Establishment of a coastal observing system

In coastal areas, physical oceanographic observations were primarily attached to biological studies. In 1931 the Institute of Marine Research (IMR) in Bergen gave Jens Eggvin the sole responsibility for building up physical oceanography in relation to fisheries and fisheries research. Physical oceanography was clearly within the range of scientific interests of the Institute, but fishery biology studies were completely dominant. It was therefore necessary for Eggvin to define the role of physical oceanography within the range of activities of the Institute of Marine Research (Sætre and Blindheim, 2002). Does physical oceanography have a value of its own or should it only be a supporting discipline for the main issues at the institute? This question often resulted in conflicts with other strong personalities at IMR, who represented the more traditional aspects of fishery biology (Schwach, 2000).

The ideas expressed in the work of Helland-Hansen and Nansen (1909) probably formed the basis for what seemed to be the guiding principles for Eggvin's work at the IMR:

- Oceanographic studies at IMR should help to explain the migration and distribution of commercial fish stocks and assist in exploring the feasibility of offering prognoses for the fishery, following approximately the same pattern as for the weather forecasts. Eggvin named this field of the oceanography as "oceanographic fisheries research" and internationally it was later referred to as "Fisheries hydrography" or "Fisheries oceanography".

In order to follow up what is believed to be the general goals of the work of Eggvin, his activity might be divided into three phases:
- Establishment of a suitable Ocean Observing System.
- Regular (annual) provision of reports on the status of the physical environment, and on special oceanographic events believed to affect the fisheries *(The diagnostic phase)*.
- Exploration of the feasibility of forecasting the impact of the physical environment and oceanographic events on the fisheries *(The prognostic phase)*.

At the Institute of Marine Research during the 1920s, the biologist Oscar Sund was in charge of oceanographic research. On his initiative the Norwegian research vessel "Johan Hjort" was fitted with a sea thermograph in 1924. In 1927 Sund also launched a measuring programme for surface temperature and salinity along the shipping routes Stavanger–Newcastle and Stavanger–Rotterdam. The temperature was directly observed and sea water samples were taken for salinity determination at IMR. At first, the observations were taken at equal intervals along the route and later at pre-designated positions. In 1931 a similar programme was started on the route from Bergen to Iceland via the Faeroes. From 1932–34 all these lines were equipped with sea thermographs supplied by Negretti & Zambra of London (Sætre and Blindheim, 2002).

Jens Eggvin of IMR continued the work of Oscar Sund, and in 1935 he initiated the installation of thermographs on board two coastal steamers, where the sensor measured the temperature of the engine cooling water intake at a depth of approximately 4 m (Sætre and Blindheim, 2002). These two vessels covered more than 2700 km of the coastal stretch between Oslo and Kirkenes near the

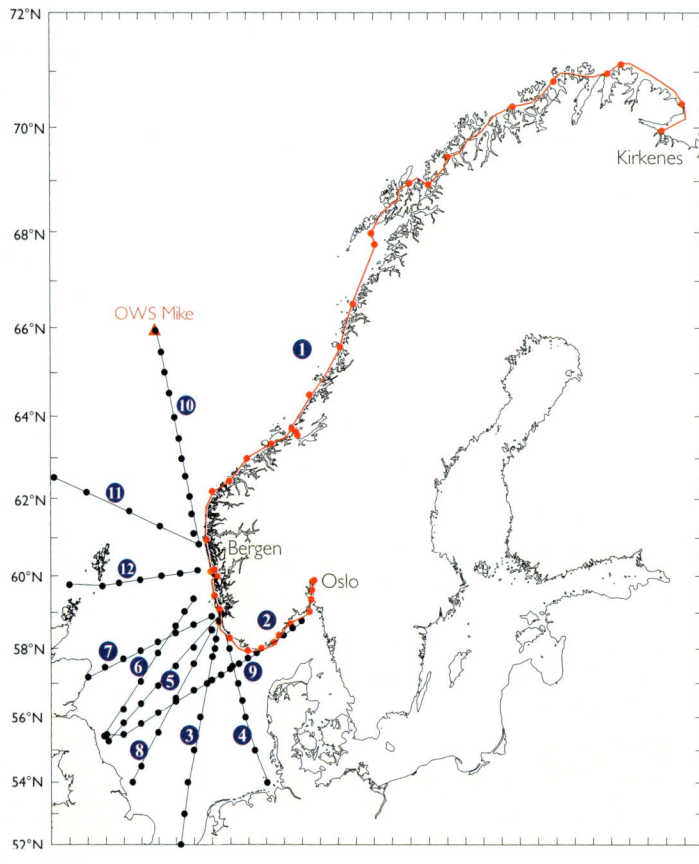

▶ **Figure 2.7**
Norwegian lines of surface observations by ships of opportunity (SOO). The names of the routes and years of operation appear in Table 2.1 (OWS Mike = Ocean Weather Station Mike).

▶ **Table 2.1**
Observation routes for surface temperature and salinity by ships of opportunity (Thermograph Service). Figure 2.7 shows the position of the observation points (Sætre and Blindheim, 2002).

Route no.	Route name	Years of operation
1	Bergen–Kirkenes	1935–present
2	Bergen–Oslo	1936–1990
3	Stavanger–Rotterdam/Amsterdam	1927–1939/1953–1991
4	Stavanger–Hamburg	1983–1988
5	Stavanger–Newcastle	1927–1939/1948–1980
6	Bergen–Newcastle	1950–1979
7	Stavanger–Aberdeen	1991–2002
8	Stavanger–Grimsby	1991–2002
9	Oslo–Newcastle	1966–1979
10	Bergen–OWS Mike	1949–1981
11a	Bergen–Iceland	1931–1940
11b	Bergen–OWS Alfa	1954–1974
12	Bergen–New York	1949–1966

▶ Figure 2.8
The first continuous record of surface temperature along the Norwegian coast 6–12 May 1935 (Eggvin, 1940).

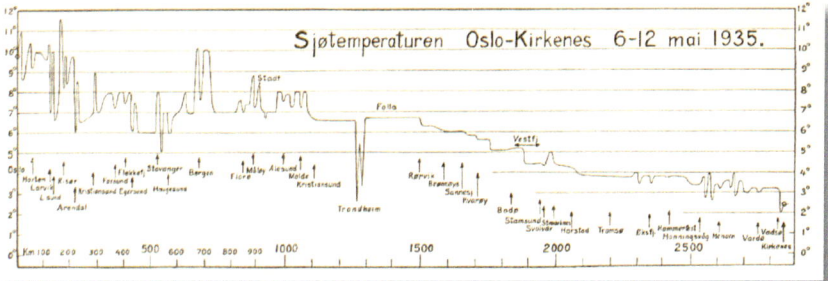

Russian border, four to eight times a month. Table 2.1 summarises all of the Norwegian observation lines by ships of opportunity and Figure 2.7 shows their position. Figure 2.8 shows the first continuous thermograph record of the upper layer temperatures on the Norwegian coast on 6–12 May 1935 (Eggvin, 1940).

On the initiative of Eggvin a network of fixed oceanographic observation stations was set up on the coast from 1935 on. These stations were located between two and eight nautical miles from the coast and were manned by local observers

BOX 2.2 THE GIGANTIC TSUNAMI THAT FLOODED THE NORWEGIAN COAST

The continental shelf edge and upper slope off central Norway north of 62°N are irregular, with very steep slopes and ridges. During the 1970s it was shown that slides of unconsolidated sediments from the edge of the shelf could have caused the slope morphology. Three main slides were recognised and the first took place more than 30,000 years ago. (Holtedahl, 1993). Based on studies of the deposits in near-shore coastal lake basins in Norway as well as observations from the Faeroes, the Shetlands and Scotland, scientists in the 1980s could identify a major slide, the Storegga slide, that took place place about 8000 years ago. This slide is probably one of the greatest submarine slides ever to take place, and the volume involved is estimated to be 2400 km³ (e.g. Bondevik et al., 2005). An earthquake approaching 7 on the Richter scale, possibly together with gas released from decomposition of gas hydrates, were probably the mechanism that triggered the slide (Holtedahl, 1993).

The Storegga slide created an enormous wave, or what today is called a tsunami. Along the coast the sea level fell to 10–20 m below the normal level in the course of less than 30 minutes.

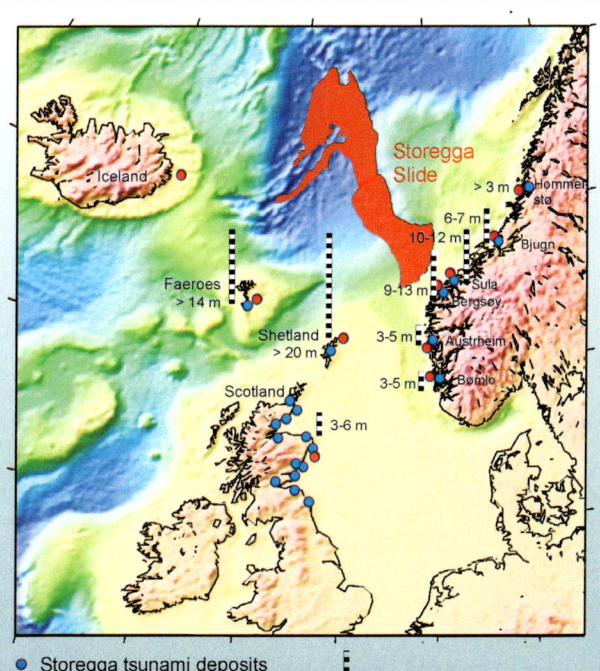

● Storegga tsunami deposits
● Location of simulated time series
┇ Run-up of tsunami deposits

▶ The Storegga slide off the coast of central Norway 8000 years ago. The figure indicate the estimated rise in sea level during the subsequent tsunami (Bondevik et al., 2005).

Then the water returned, and a monster wave of 10–12 m flooded the coastal area. Inside the fjords the wave was forced higher, and calculations suggest that the water level in extreme cases rose 40–50 m above its normal level. We have no indications of how many stone-age people were killed by the tsunami, but it must have been several thousand.

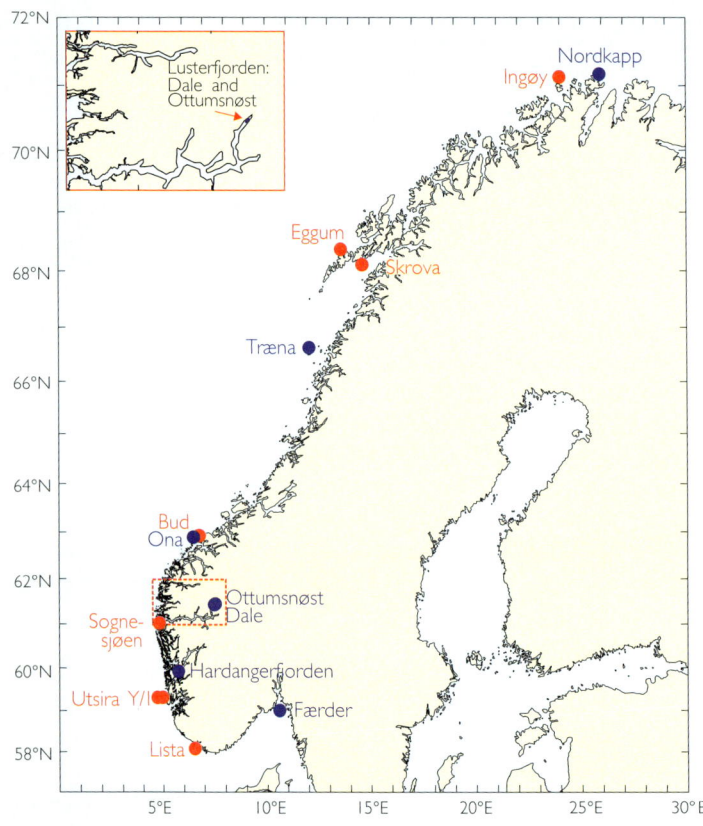

▶ **Figure 2.9**
Locations of fixed hydrographic coastal stations. Their periods of operation appear in Table 2.2. The red dots indicate those still in operation.

▶ **Table 2.2**
Names and years of operation for the fixed hydrographic stations on the Norwegian coast. Figure 2.9 shows the position of the observation points (Sætre and Blindheim, 2002).

Station name	Years of operation	Number of stations until August 2002
Nordkapp	1955–1967	174
Ingøy	1936–1944/1967–present	1180
Eggum	1935–present	1605
Skrova	1935–present	2942
Træna	1945–1948	53
Ona	1946–1954	201
Bud	1971–present	802
Sognesjøen	1935–present	1660
Otumnøst/Dale	1950–1954	178
Hardangerfjord	1955–1958	67
Utsira Ytre/Indre	1942–present	3345
Lista	1942–present	1572
Ferder	1964–1967	40

who were equipped and trained by the staff at the Institute of Marine Research (Figure 2.9, Table 2.2). They measured the temperature by means of reversing thermometers and took samples for salinity determinations using Nansen bottles at pre-selected depths, down to 200–300 m. These measurements were supposed to be carried out weekly or every second week throughout the year. The salinity samples were sent to IMR for titration. The observations from some of these fixed stations represent the longest continuous oceanographic time series in the world. It is remarkable that even during World War II, when most other European oceanographic time series were interrupted, the observations at some key stations on the Norwegian coast continued (Aure and Østensen, 1993, Sætre and Blindheim, 2002).

A network of fixed hydrographic sections running approximately perpendicularly to the Norwegian coast to the open ocean was set up during the 1930s. Some of these sections are still running and are usually repeated two to eight times a year.

Oceanographic observations were an integral part of the IMR's investigations during the seasonal fisheries for Northeast Arctic cod off Northern Norway in the winter and spring. On the basis of these observations, Eggvin drew up annual reports on the physical conditions of the fishing grounds of the cod and later also included a similar description of conditions on the fishing grounds for the Norwegian spring-spawning herring off central and western Norway. In his descriptions he attempted to explore how physical conditions influenced the distribution and availability of the fish, and these observations were published in the official Annual Report on Norwegian Fisheries. His first report dealt with the winters of 1931 and 1932 (Eggvin, 1932) and this activity was repeated every year for several decades (e.g. Eggvin, 1936; 1938).

Eggvin (1940) describes a situation from the cold winter of 1937, where he followed an outbreak of cold water and its propagation from the Skagerrak along the coast of southern and western Norway. When the cold water flushed the traditional fishing ground the herring disappeared and probably moved farther from the coast and towards deeper waters. Another effect of cold winters is the great exchange of water masses on the Norwegian coast in 1940 (Eggvin, 1943). The extreme cooling of the surface layer during the winter of 1940 resulted in the renewing of the bottom and deep water on the shelf area and in the fjords. This situation continued in 1941 and 1942, while in 1943–1944 large amounts of relatively warm Atlantic water approached the coast and the oceanographic situation from the last part of the 1930s was restored. It seems that these wide physical fluctuations had a pronounced effect on the fisheries, such as the fishery for young herring.

In recent years the term "Operational oceanography" has become quite common. The following definition has been widely accepted by the scientific community:

"Operational oceanography is the activity of routinely making, disseminating, and interpreting measurements of the seas and the atmosphere so as to:
- Provide continuous forecasts of the future condition of the sea, as far ahead as possible *(Forecasts)*.
- Provide the most usefully accurate description of the present state of the sea, including its living resources *(Nowcast)*.
- Assemble climatic long-term data sets which will provide data for descriptions of past states, and time series showing trends and changes *(Hindcast)*."

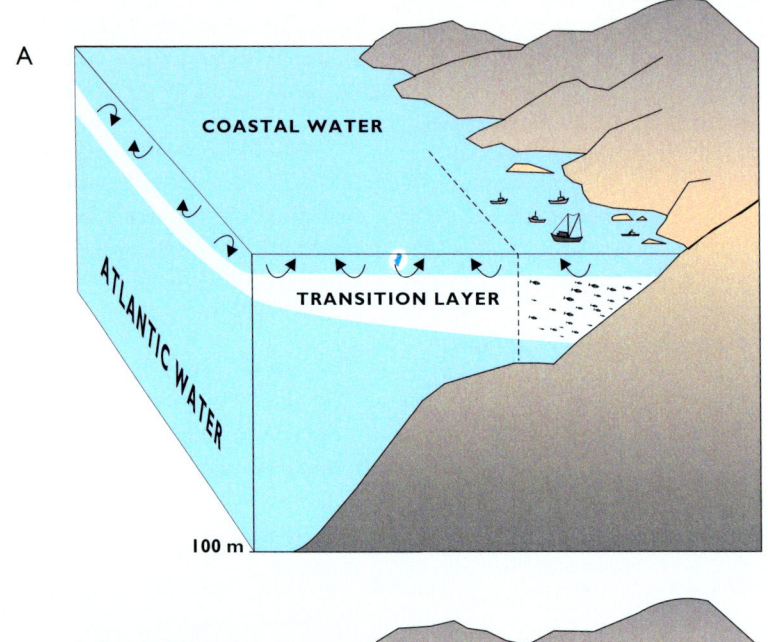

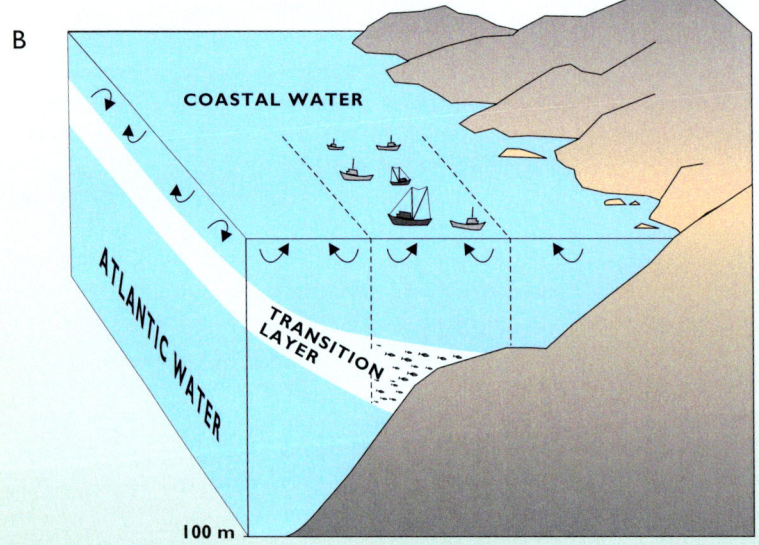

► **Figure 2.10**
Conceptual illustration of how the depth of the transition layer affects the fishery in the Lofoten area.
A) During a deep transition layer.
B) During a shallow transition layer (Modified after Sverdrup, 1952).

It was early realised that the cod in the Lofoten area preferred to spawn in the transition layer between the cold coastal water and the warmer water of Atlantic origin below. The temperature of the transition layer was between 4° and 6 °C, and the depth of this layer displayed wide inter-annual variations. The depth of the transition layer would influence both the depth and the distance from the coast of the fishery. If the transition layer was shallow the fisheries would be better in the protected and less exposed eastern part of the Lofoten archipelago and thereby available for smaller, partly open boats. During a deep transition layer situation the fishery will mainly take place in deeper waters and in the more distant and exposed western part of Lofoten (Eggvin, 1946). For many years, Eggvin submitted regular forecasts of the depth of the transition layer before the start of the fisheries (Figure 2.10).

2.3 Studies after the World War II

In the years following the Second World War most Norwegian physical oceanography concentrated on the Norwegian Sea and the Atlantic inflow. In coastal areas most marine studies were of biological character, while the physical observations were usually regarded as auxiliary data.

In June–July 1966 the Joint Skagerrak Expedition under the auspices of the International Council for the Exploration of the Sea (ICES) was carried out with participation from Norway, Sweden, Finland, Germany and Scotland. The main objective of this exercise was to study the short-term variability of the Skagerrak and Kattegat (Anon, 1970). For that purpose it was essential to map the whole area in a synoptic way. The study was repeated three times in order to allow for the influence of variable external forces.

This expedition initiated a system of fixed hydrographic sections along the Norwegian North Sea coast carried out by IMR during the summer, and Ljøen (1980) summarised the results of these for the period 1967–1976.

The Institute of Marine Research deployed close to one thousand drift bottles in Norwegian coastal waters during 1960–1970. The result obtained by these bottles are summarised by Sætre (1976; 1981). During studies of the drift of cod eggs and larvae from the spawning fields off Northern Norway during the period 1967–1973 several hundred drift bottles were deployed each year during the spring, and the results were reported annually (e.g. Hognestad, 1973).

Hydroelectric power plants modify the natural seasonal cycle of freshwater runoff to the fjords. Scientific studies of the physical and biological effects of such regulation of the fresh water started in 1969. Some of the scientists working on these projects gradually become more concerned about the potentially wider impact of freshwater regulation. Because of this concern, the Association of Norwegian Oceanographers organised a symposium in 1974 with the aim

---

BOX 2.3 THE CORIOLIS TANK – A HYDRODYNAMIC LABORATORY

► The rotating Coriolis tank at SINTEF in Trondheim.

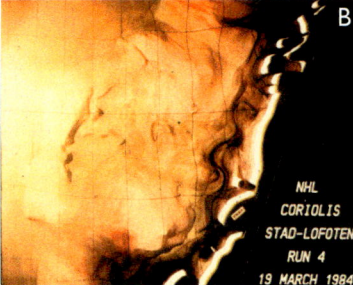

► From a run of the Coriolis tank simulating the Norwegian Coastal Current off central Norway.

A laboratory model, the rotating Coriolis tank in Trondheim, has been a useful tool (Figure A) both for practical studies and for improving strategies for developing relevant numerical models. The tank has a diameter of 5 m and a depth of 50 cm. It has been used to simulate the transport and spreading of properties of ocean currents on the entire Norwegian continental shelf including the Norwegian Coastal Current, as well as in several fjords. Three independent sources of water with different salinities allow the simulation of stratified, rotational geophysical flows. Rotation periods can be set between 10 seconds and 27 hours. The influence of the bottom topography and the establishment of stationary and transient eddies were important features of running this model for the Norwegian Coastal Current. (James and McClimans, 1983, McClimans and Johannessen, 1998).

Figure B is from a run of the Coriolis tank for the Norwegian Coastal Current off central Norway. A coastal jet close to the shore is demonstrated by the black colour. The red colour was an attempt to see how a simulated oil spill on a herring spawning ground could be spread and transported by the coastal current.

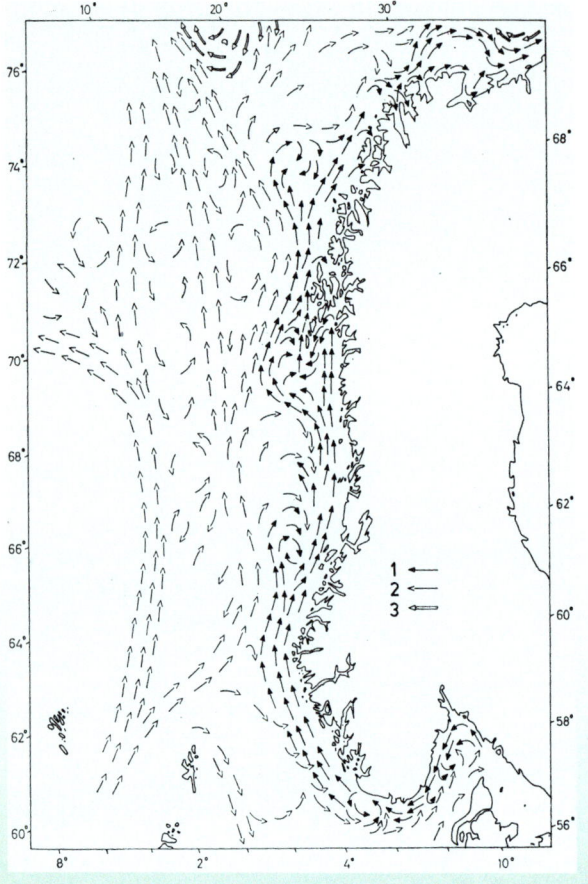

► **Figure 2.11**
Water masses and currents off Norway.
1) Coastal water
2) Atlantic water
3) Polar water
(Sætre and Ljøen, 1972).

of elucidating the influence of fresh water outflow on physical and biological processes in fjords and coastal waters (Skreslet et al., 1976). Changes in these processes resulting from alterations in the freshwater runoff could be observed in some of the fjords, while it was difficult to find any indications of such an impact in more open coastal waters. However, there are still several unanswered questions regarding whether and how the variability in the freshwater runoff from Norway influences physical and biological conditions in the Norwegian Coastal Current.

Besides freshwater regulation, the development of Norwegian petroleum exploration and marine aquaculture also contributed to the expansion of physical oceanographic research in coastal areas in the early 1970s. Greater government funding of marine sciences, together with improved cooperation between physical oceanographers and marine biologists, support the idea that the 1970s represented the start of a new golden age for coastal research. In the course of the past 30 years there has been a gradual increase in the number of scientific publications from coastal areas.

Sætre and Ljøen (1972) published the first current map with specific emphasis on the Norwegian Coastal Current (Figure 2.11). It was a composite of all available observations of water mass distribution and current measurements. In 1983, the map was updated to take into account more recent knowledge (Figures 2.12a, b and c). The "Norwegian Coastal Current" project was launched in 1975 on the initiative of the Norwegian Oceanographic Committee. The main objectives

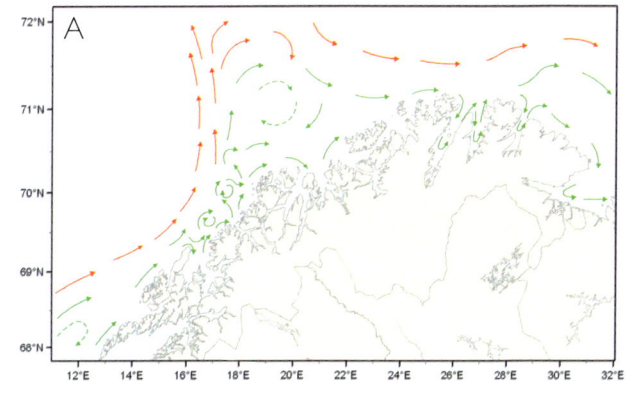
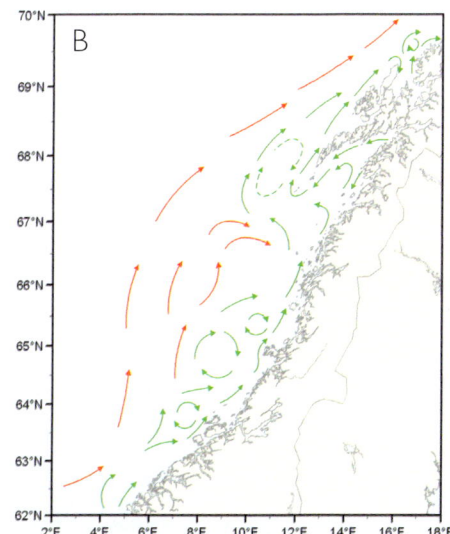
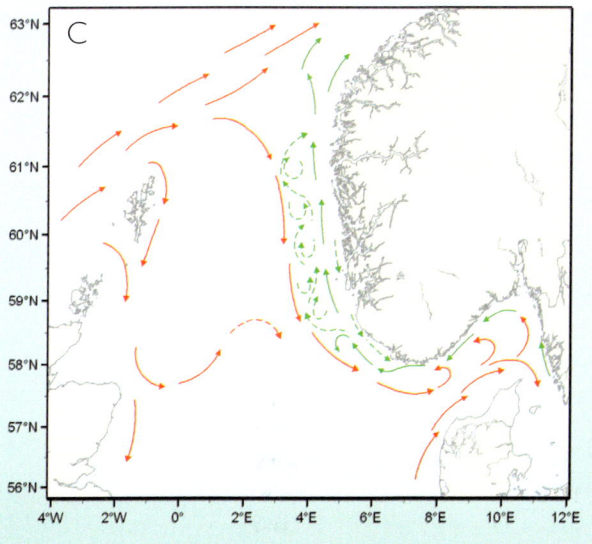

▶ **Figure 2.12**
Mean currents in the surface layer off the northern (A), central (B) and southern (C) coast of Norway. Red arrows indicate water of Atlantic origin and green arrows the coastal water (Sætre, 1983).

of the project were to acquire a better basic understanding of oceanographic processes in the coastal current, such as structure, dynamics, currents and transport. One of the first tasks of the project was to produce a list of all the available scientific literature on the physical, chemical, biological and geological oceanography of Norwegian coastal waters (Anon, 1975). The field phase started with a synoptic oceanographic study covering the whole of the Norwegian coast during the period 24 May – 3 June 1975 (Anon, 1976). The project was formally terminated in 1980 and the main results were presented at a symposium the same year (Sætre and Mork, 1981). The project benefited from wide international cooperation, such as with the "Joint North Sea Data Acquisition Programme 1976" (JONSDAP '76) and the Norwegian Remote Sensing Experiment (NORSEX) in 1979 (Anon, 1979).

In the course of the past few decades, numerical models have developed into powerful tools for oceanic and coastal research. Ocean circulation models are used to describe physical dynamics and states, and incorporate such variables as temperature, salinity, currents and sea level, and their temporal and seasonal fluctuations. These models are based on physical laws, where the associated equations cannot be solved analytically. Therefore, simplified versions of the equations are solved numerically in a predetermined grid. The associated calculations are very extensive and demand considerable computing power.

Numerical oceanographic modelling in Norway started in the mid-1970s, with the development of a model for prediction of sea level elevation caused by wind and atmospheric pressure. Since then, numerical modelling competence has developed at a number of Norwegian scientific institutions and research groups. Today, the integrated family of models consists of general ocean circulation models, various transport models and models for the status and development of phytoplankton and zooplankton stocks. The quality of the results obtained by these models has improved considerably in recent years. The most important bottleneck for coastal research has been the horizontal resolution or grid size. A coarse grid will miss finer topographic details and smaller features, such as fronts and eddies, while a finer grid may give more realistic results but demands much higher computing capacity. Fine-grid models therefore usually cover a smaller geographical area.

During the 1980s numerical models for the coastal region became available, and these useful tools have gradually been improved. In 1990 work started on a coupled physical-chemical-biological model for the Nordic Sea as a co-operation between the Institute of Marine Research, the University of Bergen and the Norwegian Meteorological Institute. The basis is a sophisticated three-dimensional physical model and the total model system is now referred to as the NORWegian ECOlogical Model system (NORWECOM) (Skogen, 1993).

A number of conspicuous ecological events took place in the Barents Sea ecosystem and in the coastal waters of northern Norway during the 1980s. This included decimation of the kelp forests, the collapse of the capelin stock with severe consequences for cod stocks, and a massive invasion by harp seals searching for food. The Norwegian Research Programme on North Norwegian Coastal Ecology (MARE NOR) was launched in 1990 in response to the need of the national authorities to better understand the abrupt and marked variations in the coastal ecosystem of northern Norway. One of the first tasks of MARE NOR was to collect all the available literature of relevance for the Norwegian coastal ecosystem (Dalpadado, 1989). MARE NOR came to an end in late 1994 and a concluding international symposium was organised in the same year (Skjoldal *et al.*, 1995). As a follow up of the MARE NOR activities a small popular scientific book in Norwegian describing the main feature of the coastal ecosystem, including the physical environment, was published (Rinde *et al.*, 1998).

Between 1988 and 1994 a number of Nordic and Baltic scientists carried out a study of physical and chemical conditions in the Skagerrak (SKAGEX). In the main phase 1990–91 the experiment consisted of four multi-ship surveys named SKAGEX 1 through 4. The first one was by far the most extensive, with 17 research vessels and a duration of one month in May–June 1990. All in all about 200 scientists have been involved to a greater or lesser extent. The Nordic Council of Ministers largely funded the programme. The main results of the programme and a description of the work done can be found in Dybern *et al.* (1994). The report also includes a presentation of the SKAGEX Atlas, which contains a wealth of data collected during the programme.

During the past decade Norwegian scientists have put a great deal of effort into climatic programmes, both national and international. A recent programme with a more coastal approach is entitled Monitoring the Norwegian Coastal Zone Environment (MONCOZE). This programme will come to an end in 2005 and its overall objective is to develop, test and demonstrate a pilot system for monitoring and predicting the physical-biochemical conditions of the Norwegian marine coastal environment and its open boundaries. This is actually carrying further the vision of Jens Eggvin seventy years ago, but the powerful scientific tools available today make it much more likely to succeed.

The origin of the coastal zone

Terje Thorsnes and Oddvar Longva

3

3.1 The coast – living ground for the first Norwegians

The first Norwegians arrived in Norway around 11,000–11,500 years ago (Bang-Andersen, 2003). They met a cold and barren coast, but with rich opportunities for hunting and fishing. The landscape they saw was different from what we see today. Thick ice sheets covered the central mountains, and the valleys were filled with glaciers reaching down to the sea, fairly similar to what we see in Svalbard today (Figure 3.1). Thick ice sheets had covered Norway for long periods during the previous 100,000 years (Figure 3.2), and the weight of the ice had pressed large parts of the mainland down – only the outer coast along western Norway had largely escaped. In the Oslo area the sea level was as much as 220 m higher than it is today; In Trondheim it was 180 m higher than at present while, in Sta-

▶ Figure 3.1
An ice-covered coast with drifting icebergs like this may have been the first view of Norway, 11,500 years ago.

Photo: Terje Thorsnes

► **Figure 3.2**
During the ice ages, Norway was covered with thick ice sheets, bulldozing material from the mountains to the shelf.

vanger it was only 25 m higher. This means that much of the present coast was submerged. Vegetation was sparse, and the land was covered with clay, sand, gravel and boulders, left over from the thick ice sheet that had covered the entire country only a few thousand years previously.

In the North Sea, the sea level 11,000 years ago was lower than it is today. Most of the sea floor was dry land and remained so until about 7000 years ago. The North Sea plateau was separated from Norway by a fjord several hundred metres deep and 70–100 kilometres wide, running from the Skagerrak to Stad – what is now known as the Norwegian Trench (Longva and Thorsnes, 1997). In the southern part of the North Sea, trawl nets have brought up man-made tools, showing that what is now the North Sea was used for hunting and fishing. Possible stone tools have been found in grab samples and cores in the North Sea off Mid-Norway (Rokoengen and Johansen, 1996). These early people made canoes and crossed the Norwegian Trench, perhaps landing at Jæren near Stavanger. Remains of early immigrants have been found just north of Stavanger, on the island of Rennesøy. The artefacts tell a story of a settlement with a strong maritime orientation and adaptation. Hunting marine mammals and fishing was important (Prøsch-Danielsen and Høgestøl, 1995). But the sea was also a deadly enemy. Around 11,200 years ago, a tsunami – a flood wave caused by a nearby submarine slide – extinguished the settlement (Prøsch-Danielsen et al., 2005). Tools and other artefacts were washed away, and were later uncovered by a team of archaeologists in 1990.

Large slides, both on land and in the fjords, were common in the first few thousand years after the ice withdrew (Vorren et al., 2006). The extensive erosion – grinding of the bedrock – and enormous rivers of meltwater had left thick deposits of sediments covering the landscape and filling in the fjords. Rapidly deposited sediments are notoriously unstable, and large-scale slides were the result.

The early immigrants slowly moved northwards, living from fishing and hunting, and exploiting geological resources. Stone tools were necessary for survival, and the Stone Age people discovered the most suitable rock materials. On Stakaneset, on the west coast of Norway, a large stone axe mine was active from 8000–9000 years ago. The mine is situated within a diabase dyke. This is a dark rock, hard, fine-grained and massive, which makes it very suitable for axe manufacture. Its properties are due to the fact that diabase is a rapidly cooled dyke rock with a dense, fine-grained structure.

The Stone Age people lived in tents or caves. These caves were the results of waves that had excavated bedrock fracture zones. The rich communities of the Fosna culture, dating back to 10,000 years ago, provides a good example of such caves. Later, the people moved towards the fjords, establishing larger communities that were dependent on agriculture. In what is now the County of Trøndelag, large, fertile fields of recently uplifted seabed were available to the early farmers.

The Stone Age people and their descendants learned to exploit a diverse and rich coastal zone. Let us take a look at why the coast is the way it is.

3.2 Coastal landscapes – the grand picture

The concept of "coast" in this book has been used in a very broad sense – extending from the innermost part of the fjords to the outer limit of the shelf, as this is roughly the area influenced by coastal waters (Figure 4.9). In order to provide a framework for the concept of "coast", we have also included a brief description of its deeper parts.

The width of the coast according to this definition varies greatly, from less than 100 to more than 300 kilometres. If we measure the distance from the outer tip of the outermost islands to the generalised outer limit of the shelf, the range is even more impressive – from 15 to 250 kilometres (Figure 3.3).

Norwegian waters comprise a wide range of environments, from the deep sea, via the continental slope and continental shelf to the inner coastal zone with its strandflats, archipelagos and fjords. This provides a geological diversity that is unique in European terms (Box 1.1).

An exciting geological history lies behind this diversity – a development that has taken place in the course of more than 400 million years. The continents consist of plates of solidified rock that float on partially molten rock, and these plates move relative to each other. Where they collide, the Earth's crust is folded and mountain ranges are created. Where they drift apart, deep oceans form and new sea floor is created along rifts because molten rock (magma) flows up from below. A good example is the Mid-Atlantic Ridge, including Iceland, which is a

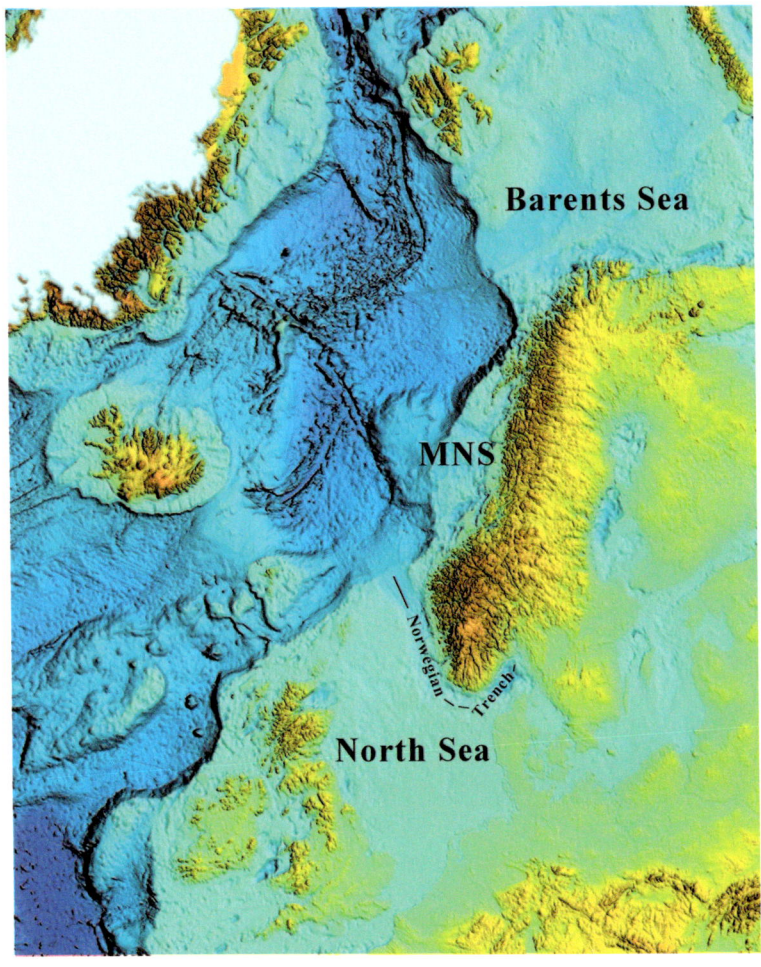

► Figure 3.3
Overview of Norway's maritime areas – from the North Sea to the Barents Sea. MNS – Mid-Norway shelf.

result of Greenland and Europe drifting away from each other at a rate of about 2 cm a year (Figure 3.4).

The tectonic plates on which Norway and Greenland rest collided more than 400 million years ago and formed mountain chains on either side of a shallow sea (Torsvik et al., 2002). Both Greenland and Norway are remnants of worn-down mountain ranges. Between these mountains, the shallow sea gradually filled up with sediments derived from erosion of the mountains. These sediments were transformed into sandstones, shales and limestones as the sediments became buried and subsided. Organic-rich sediments gave rise to coal, oil and gas. These rocks are now the source of Norway's hydrocarbon production.

The deep ocean – the Norwegian Sea – as we know it today, began forming some 60 million years ago, when the plates started drifting apart, the crust fractured and the Norwegian Sea segment of the Atlantic Ocean, with depths of more than 4000 m, began to take shape (Figure 3.4). At around the same time, the mainland was uplifted, giving rise to the Norwegian mountain chain.

Some three million years ago, the global climate became so cold that ice ages developed (Vorren and Mangerud, 2006). Since then, Norway has been glaciated many times and the glaciers have eroded the mountains, excavated valleys and fjords, and bulldozed the eroded material out to the shelf and dumped it. During each ice age, the shelf has grown, and the margin and continental slope have been

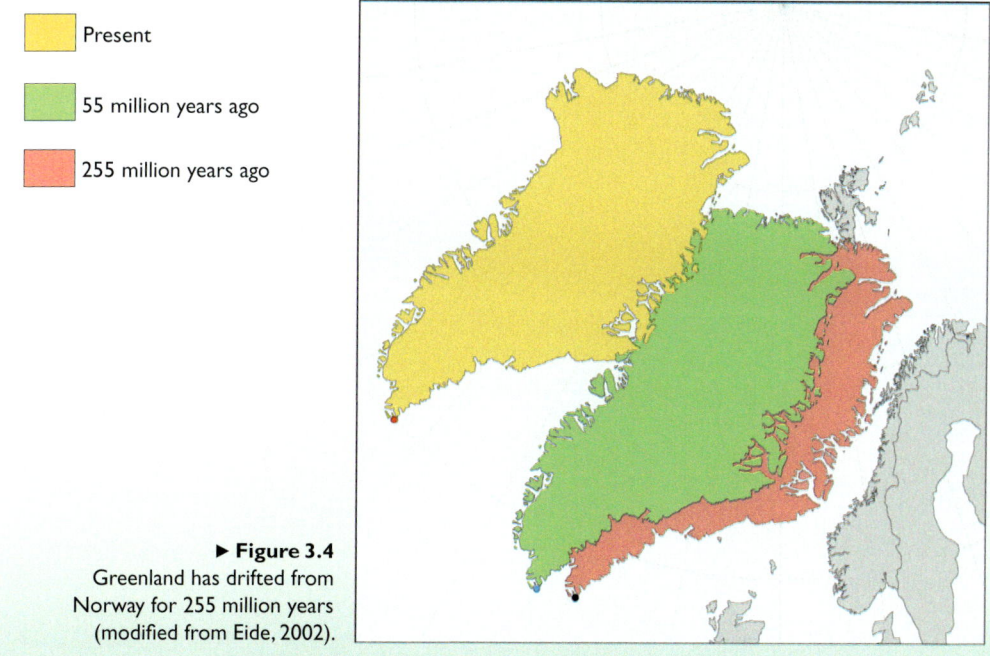

▶ Figure 3.4
Greenland has drifted from Norway for 255 million years (modified from Eide, 2002).

Present
55 million years ago
255 million years ago

moved further and further from the inner coast, particularly off central Norway between the Halten Bank and the Lofoten archipelago. Starting at approximately the same time – three million years ago – the final uplift of the Norwegian mainland took place, creating the mountains that we see today.

The continental slope comprises the area from the margin of the shelf into the deep ocean (Figure 3.5). The foot of the slope is located approximately where Norway and Greenland separated 250 million years ago. The slope is a notable submarine terrain feature (Figure 3.3) extending from the British Isles, past central and northern Norway, and all the way past Svalbard (Holtedahl, 1993). Major submarine slides have shaped parts of the slope. Northwest of Møre, the giant Storegga slide occurred 8000 years ago, creating an enormous tsunami which hit the west coast of Norway (Box 2.2). Flood waves between 10 and 15 metres high hit the coast, and must have been a devastating event. The backwall is a several hundred kilometre-long escarpment, which makes up the break in the shelf. The area is known for its rich fisheries and numerous coral reefs.

▶ Figure 3.5
Cross section from the deep ocean to the fjords.

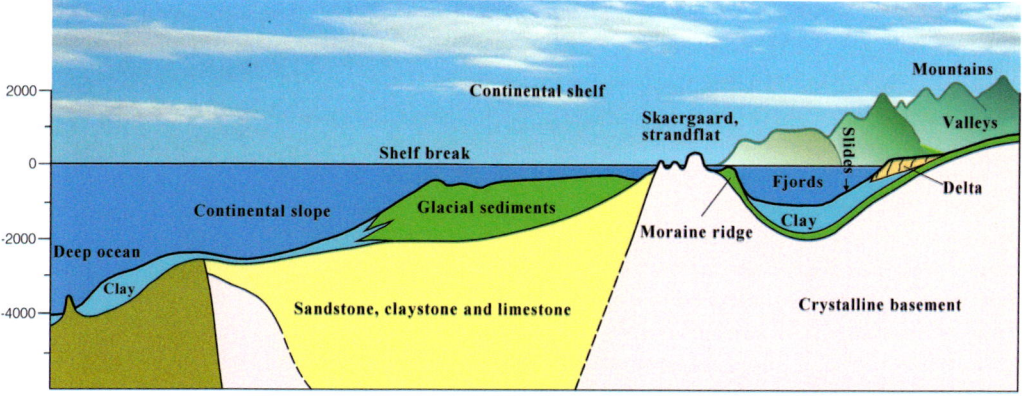

The origin of the coastal zone   39

► Figure 3.6
The Norwegian "skaergaard" with thousands of islands and skerries supports important fisheries and aquaculture production, and is the result of complex geological processes.

Photo: Turid Helle

The continental shelf is the area stretching from the margin of the shelf to the coastline. The continental shelves off Mid-Norway and North Norway are rather uneven consisting of shallow banks separated by deeper troughs. This submarine landscape is the product of the uneven flow of the glaciers on their passage from mainland Norway westwards to the margin. Glaciers flowing through troughs such as the Træna Deep were "conveyor belts" for enormous amounts of eroded material, which was transported to the edge of the shelf and dumped there. At the same time, the sheet over the bank areas such as the Træna Bank was stagnant, and left little imprint on the seabed (Ottesen et al., 2005)

The continental shelves in the central North Sea and in the central Barents Sea are smoother, with low relief. An exception in the North Sea is the Norwegian Trench, which reaches along the coastline from the Skagerrak to Stad, attaining a maximum depth of 700 m (Figures 4.9 and 3.3). This trench was excavated by ice that flowed out from south-east Norway and Sweden, coalescing to flow as an ice stream along the coast and calving in the Norwegian Sea off Stad.

The "skaergaard" is a typical feature of the Norwegian coast (Holtedahl, 1993). This archipelago with islands and skerries (Figure 3.6) forms part of the "strandflat", an area of irregular terrain stretching from 40 m below to 40 m above sea level, with the exception of some deeper channels. It forms a rim of varying width around much of Norway, with a multitude of islands and islets, which are particularly widespread on the southern coast and between Stavanger and Lofoten. The deeper depressions run parallel to the coast, or at right angles to it, and follow fractures and fault zones, or zones of softer bedrock. The strandflat was formed by frost action and by marine and glacial erosion. A typical feature of the strandflat is the presence of islands composed of a flat brim encircling higher ground, which are the remains of a former more elevated landscape.

The fjord mouths are located where the mountainous hinterland opens out onto the strandflat. The fjords have a shallow threshold or sill at their mouth and are deeper further in (otherwise they are correctly termed sea bights). The fjords were excavated by glaciers along zones of weakness (fracture and fault zones) in the bedrock and became deeper with each glaciation. Many of the valleys date back as much as 150 million years – to the Jurassic period.

3.3 Conditions on the sea floor – from "Eggakanten" to the fjords

The continental shelf extends westwards to the shelf break ("eggakanten"). On the continental slope, outwards of the shelf, lie clayey sediments. Powerful currents usually flow along the break, the finest material is winnowed, and the sea floor consists of rocks, gravel or sand. Coldwater coral reefs are found on long stretches of the shelf margin, especially along Storegga and the Træna Deep. In these areas, deep slide scars can be seen, the currents are strong, and the bottom offers substrates on which the corals have been able to gain a foothold. Ploughmarks made by icebergs drifting in the ocean more than 10,000 years ago provide a special environment for coral reefs in many places.

▶ **Figure 3.7**
"Eggakanten" is the Norwegian term for the break between continental shelf and continental slope. Extensive submarine sliding has locally produced dramatic terrains, hosting extensive cold water reef communities. Water depths between 290 m (right) and 570 m (left). See figure for scale.

The continental shelf can be divided into three broad regions – the North Sea, the shelf off Mid-Norway and north to Finnmark, and the Barents Sea (Figure 3.3). In the North Sea, the sea floor is relatively even, and consists mostly of sand and/or gravel, except in the Norwegian Trench where a clay-covered floor is found in the deeper parts. The shelf off Mid-Norway and north to Finnmark is uneven, with troughs and banks. Many of the banks were either dry land or were very shallow for a brief period at the end of the Ice Age. As a result, the sea floor often consists of a mixture of gravel, sand and mud, just as we can see in the present shore zone. The most fine-grained material – clay – was winnowed and deposited in the troughs to form muddy bottoms. Powerful local currents sweep along the floor of the troughs. In these places, there is no deposition of fine-grained material, and we find a mixture of sediments, left by the ice sheets that

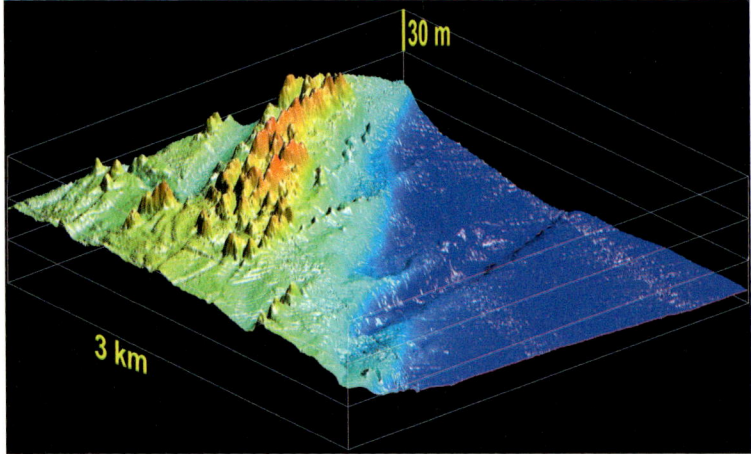

▶ Figure 3.8
3D model of the Sula Reef with numerous coral mounds, based on multibeam bathymetry. Water depths range from 275 to 325 m, vertical exaggeration c. 10x.

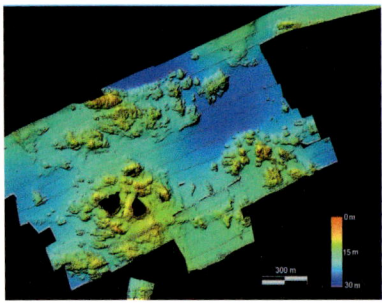

▶ Figure 3.9
Typical skaergaard submarine landscape, with rock knolls separated by basins with sand and mud.

crossed the shelf more than 12,000 years ago. The banks are important spawning areas for both demersal and pelagic species of fish. Generally speaking the seafloor in shallow parts of the Barents Sea is a mixture of gravel, sand and mud, while the deeper parts like the Bear Island Channel are covered by clayey sediments.

Gas and fluids leak from the bedrock and/or sediments on many parts of the shelf, and reduce the sedimentation at the seabed, occasionally causing blowouts. These processes form hollows in the seabed known as pockmarks. They are usually 50–100 m wide and 4–10 m deep, but in some places, such as the southern slope of the Norwegian Trench in the Skagerrak, there are pockmarks which are 1 km long and up to 30 m deep. Bacteria and other organisms have been found in the pockmarks, where they live on the gas and fluids, and attract fish which graze there.

Cold-water coral reefs are also found on the shelf. The best known of these is the Sula Reef (Figure 3.8), which lies 50 km north of the island of Frøya, off the coast of Trøndelag. This reef complex is composed of cones of stone coral (*Lophelia pertusa*) as much as 30 m in height, situated on the crest of a 15 km-long ridge of bedrock. The coral reefs house a wide range of life forms, and the adolescent stages of many of the species of fish that are commercially exploited in Norway grow up in these areas.

In the coastal zone, the bottom consists of alternating rocky knolls and channels whose floor is covered in a variety of sediments, clay and mud, sand, stones and gravel (Figure 3.9). In the transitional zone from breaking waves to calmer water, or in straits with powerful currents, large quantities of shell-sand often accumulate. The shallow water gives good light conditions, and bedrock and stones offer footholds for the large fronds of sea tangles which form kelp forests. This alternation of different types of bottom produces an enormous abundance of different species of fish and crustaceans.

The fjords are much deeper than the coastal waters. Their outer boundaries are marked by shallow thresholds, separating the deep fjords from the shallower archipelago. Their sides tend to take the form of steep cliffs, between which is a flat bottom covered with clay and mud. In the transition zone between the bottom and the sides of the fjord there are often avalanche fans formed by rock and snow slides and rock falls, or gravel and sand deposits carried down by rivers and streams. Where valleys enter the fjord, the rivers flowing into the fjord generally build up large deltaic deposits composed of gravel and sand. Many fjords have large or small moraine ridges with a rugged surface, and these form local thresholds across the fjord. The ridges were formed at the glacier terminus during the deglaciation when the glaciers made temporary ad-

▶ **Figure 3.10**
Overview from the Trondheimsfjord to Haltenbanken. Note the extensive skaergaard (S) around the islands, forming part of the strandflat, and the thresholds (T) in the outer parts of the fjords.

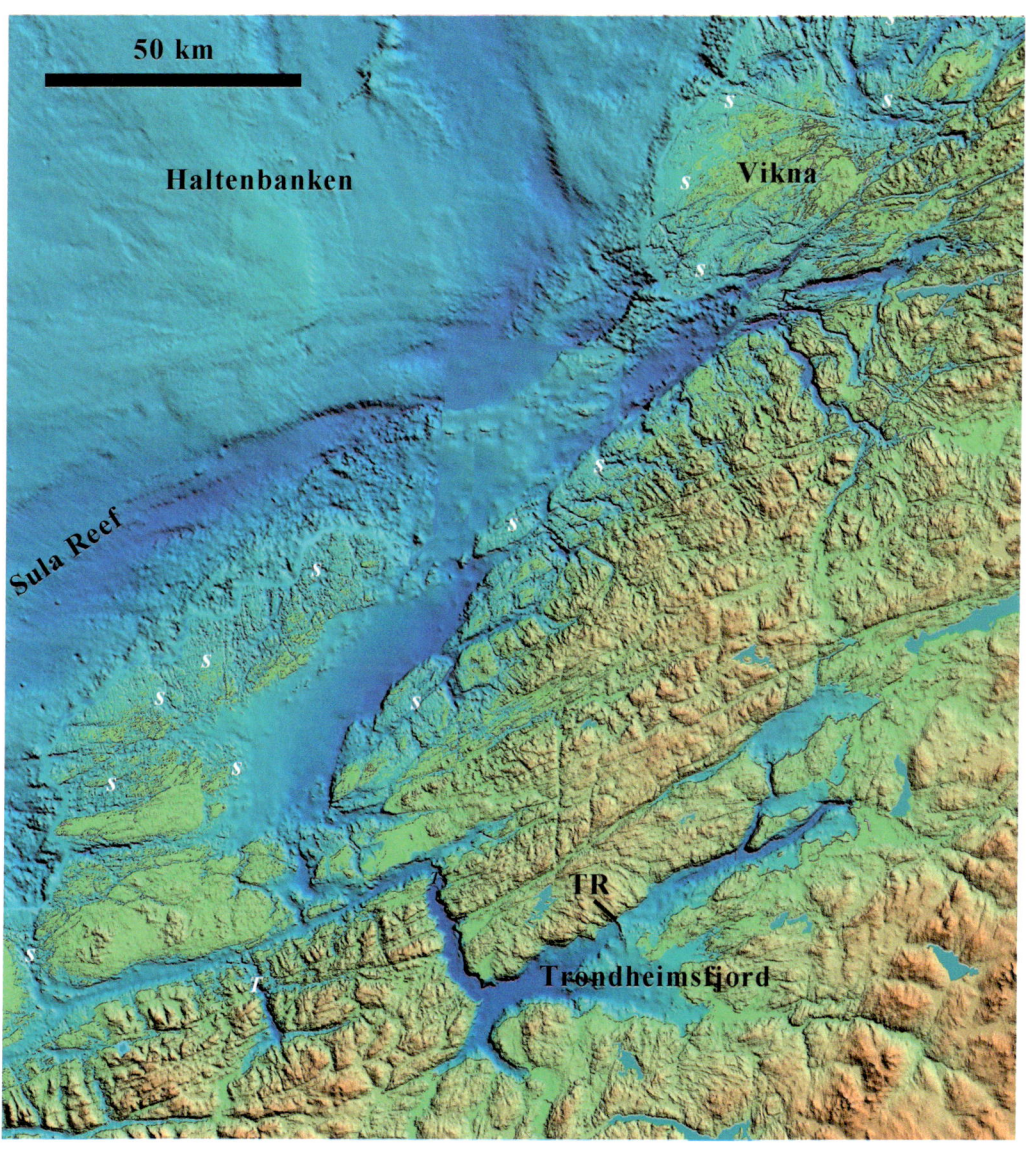

The origin of the coastal zone 43

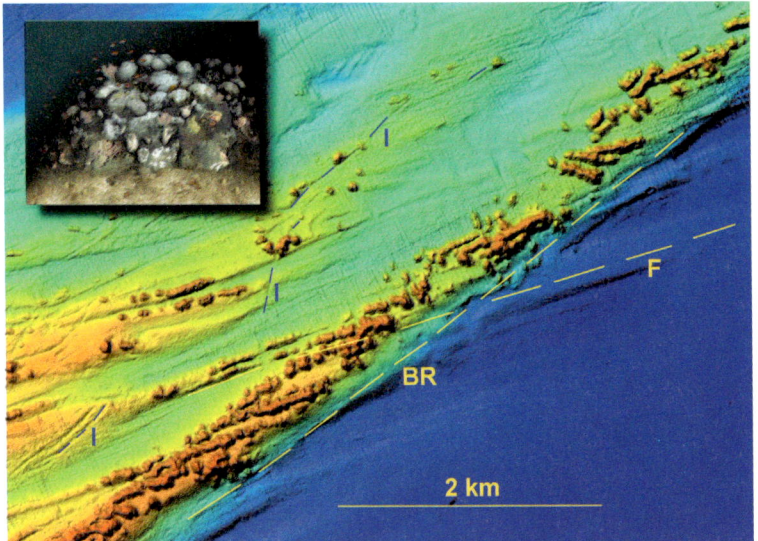

► Figure 3.11
Detail from the northern part of the Sula Reef. A prominent bedrock ridge (BR) crosses from lower left to upper right. Flutes (F) cross the bedrock ridge at ca. 20 degrees angle. Iceberg plough marks (I) appear as irregular depressions. Coral mounds (red to brown elevated features) occur mainly along the bedrock ridge. Water depths between 275 and 325 m. Inset photo – a coral mound montage (Source: Institute of Marine Research).

vances or halted for shorter or longer periods during its retreat. The Tautra Ridge (TR) in Trondheimsfjord is one such moraine ridge (Figure 3.10). Selligrunnen, a shallow area with many coral reefs, is located on the Tautra Ridge.

3.4 Geodiversity = biodiversity

The Norwegian oceans are exceptional in many ways. They support some of the world's richest fisheries, they contain huge hydrocarbon resources, and if we include the onshore region, have enormous total resources. The coast is Norway's economic backbone, and is fundamental to the nation's prosperity. The coast also provides the framework for a very rich biodiversity, ranging from primitive bacterial communities in the sediments, through coral reefs and sponge communities, to pelagic fish and mammals. This is linked to the wide geodiversity of the coast – a multitude of rock types ranging from gneisses to shales, sediment types ranging from moraine to mud, and processes ranging from crustal uplift of mainland Norway to iceberg ploughmarking all contribute to the physical diversity that is fundamental to supporting some of the world's richest ecosystems.

One of the most striking examples is the Sula Reef on the Halten Bank (Figure 3.11, see Figure 3.10 for its location). The Sula Reef is a 13 km-long, 400 m-wide complex, with more than 500 coral mounds. The mounds are either solitary or grown together in ridges. The majority are found along the Sula Ridge – a bedrock ridge formed by a resistant sandstone layer. Looking in greater detail at the distribution of the mounds, shaded relief images based on multibeam bathymetry reveal an en echelon arrangement of the mounds relative to the ridge. The mounds grow selectively on top of low sediment ridges which cross the bedrock ridge. These sediment ridges (flutes) were created beneath a major ice sheet that extended into the shelf area more than 10,000 years ago.

The wide geodiversity of the coast of Norway is the result of a long and dramatic past – and it supports present-day and future biodiversity.

Driving forces

Roald Sætre

Regions of the coastal seas are affected both locally and remotely by oceanic, atmospheric, bottom and terrestrial interactions. The following factors influence the general features of movements and properties of the Norwegian Coastal Current in both the short and the long term:
- Freshwater
- Tides
- Wind conditions
- Atlantic water
- Bottom topography

4.1 The influence of freshwater
There are three sources of freshwater in the Norwegian Coastal Current, which contribute to the total approximately as follows:

Baltic outflow	50 %
Freshwater runoff from Norway	40 %
Freshwater runoff to the North Sea	10 %

The Baltic outflow and the freshwater runoff to the North Sea contribute significantly to the load of pollution and nutrients in the Norwegian coastal region.

4.1.1 The Baltic outflow
The classical estimate of the annual freshwater supply to the Baltic is 470 km^3/year or about 15,000 m^3/s (Knudsen, 1899). This is mainly determined by river input, as evaporation and precipitation during an average year will approximately balance out. More recent calculations estimated an average total long-term net freshwater supply of about 500 km^3/year or 15,700 m^3/s, which is not far from the classical value. This freshwater mixes with seawater to form an brackish water outflow of 1,100 to 1,300 km^3/year, equivalent to 35 to 40,000 m^3/s There are large short-term fluctuations in the outflow and the instantaneous flux of water quite often exceeds 100,000 m^3/s, with maximum fluxes of more than 200,000 m^3/s. This temporal variability is driven by differences in the sea level of the Kattegat and the South-Western Baltic proper, which in turn is determined by spatial air pressure and wind variations (e.g. Rydberg, 1996). The amount of water in the Baltic Sea varies by some 370 km^3 on a time scale of a few weeks, corresponding to about 1 m in sea level. The air pressure differences can be used to calculate the volume flux with surprisingly high accuracy. A rise in the level of the Baltic Sea of 1 cm/24 h is equivalent to a barotrophic transport of 50,000 m^3/s through the

Kattegat. Typical variations are of the order of 5 cm/24 h, equivalent to a mean current speed through the Kattegat of 10 cm/s (Svansson, 1984).

The average seasonal variations in the Baltic outflow appear to be a complex function of the annual fluctuations in the freshwater supply to the Baltic and the seasonal cycle in sea level and thereby the variations in the volume of the Baltic. As a consequence of this complexity, the outflow is completely out of phase with the freshwater supply to the Baltic. Various attempts to calculate the mean seasonal variation in the Baltic outflow show some discrepancies. It has previously been suggested that the maximum outflow occurs during January–May. The lowest average value is observed in June and July, following which there is an increase in August (Wyrtki, 1954; Svansson, 1975). According to Gustafsson (1997) the outflow has a maximum in January–February and two pronounced minima, one in March and another in June. The flow across the border with the Skagerrak varies, as might be expected, more smoothly, with a maximum in February–March (around 22,000 m^3/s) and a minimum in July (around 11,000 m^3/s).

If we consider the Skagerrak only, this area receives an average of around 2,100 m^3/s fresh water, of which approximately 75 % is from the Baltic outflow, 15 % from the North Sea and 10 % is local runoff from Sweden and Norway. Obviously, given these figures, the Baltic outflow must have a significant effect on the coastal current along the southern coast of Norway. The seasonal pattern for the freshwater runoff from Norway deviates significantly from that of continental Europe and the runoff to the Kattegat. Its maximum discharge is during late spring and summer, while the minimum flow typically occurs from December to March.

As far as long-term variability is concerned, the Baltic Sea is nearly enclosed and the water balance does not average out on an annual or even decadal scale. The annual average sea level may deviate by 10 cm from the long-term mean (Kahma *et al.*, 2003). Attempts to correlate the sea level variations of the Baltic with local atmospheric conditions have not been very successful. On a larger scale, however, a correlation between water levels and atmospheric pressure patterns described by the North Atlantic Oscillation (NAO) can be observed (Heyen *et al.*, 1996). The NAO index is determined from the difference in atmospheric sea level pressure between the Azores high and the Icelandic low and is a measure of the strength of the westerly flow. Kahma *et al.* (1996) demonstrated a high positive correlation ($r=0.8$) between the 15-year running mean of the NAO winter index and the residual water level on the Finnish coast. On an annual time-scale, the correlation was reduced to $r=0.6$ but was statistically more significant because of the higher number of degrees of freedom.

4.1.2 Freshwater from the North Sea
Total freshwater runoff to the North Sea for the period 1980–1990 has varied between 3,500 and 6,000 m^3/s. The Rhine and the Meuse contribute the largest flows, while the Elbe has the second largest single discharge. Between 1980 and 1990 these rivers contributed an average of 3,877 m^3/s. The freshwater flow to the German Bight is high in winter, with a peak in March/April followed by a decrease in late spring with a minimum in late summer/early fall (Baliño, 1993). It is not clear how much of this water is really transported into the Skagerrak by the Jutland Current. Estimates based on hydrographic measurements indicate between 30 % (Rydberg *et al.*, 1996) and 50 % (Rohde, 1989). A reasonable "guesstimate" used in some assessment work seems to be around 2,000 m^3/s.

▶ Figure 4.1
Runoff regions in Norway. Modified and redrawn after Tolland (1976).

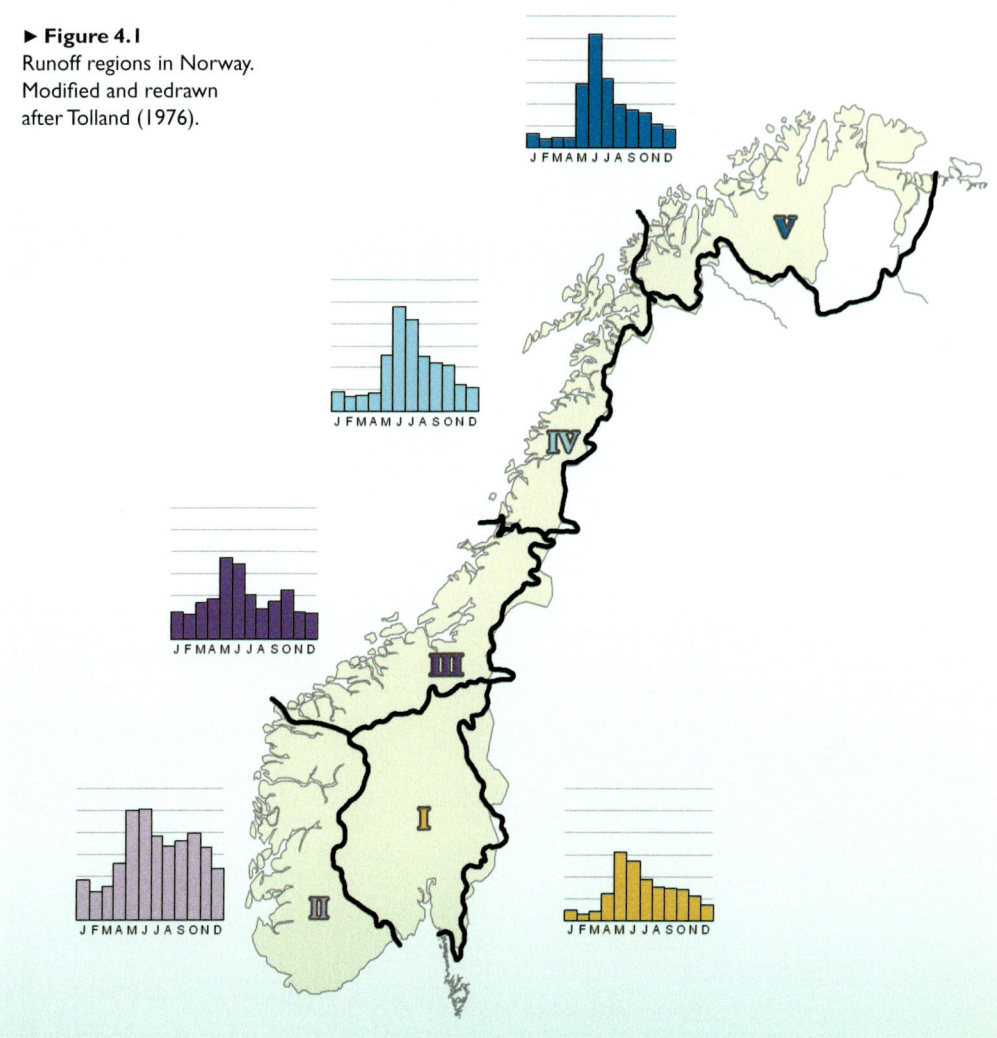

▶ Table 4.1
Normal monthly natural runoff in m³/s from Norway by region (Figure 4.1), 1931–1960 (Recalculated after Tollan, 1976).

Region	Jan	Feb	Mar	Apr	May	Jun	Jul	Aug	Sep	Oct	Nov	Dec	Year
I	47	32	41	120	309	268	186	154	145	139	110	72	1623
II	183	132	151	261	501	507	381	334	359	397	328	233	3767
III	123	113	164	183	369	340	198	135	173	224	126	120	2268
IV	88	66	72	82	258	473	416	252	221	205	120	107	2360
V	66	41	47	47	290	517	315	195	170	158	104	82	2032
Total	507	384	475	693	1727	2105	1496	1070	1068	1123	788	614	12050

4.1.3 Freshwater runoff from Norway

The total runoff from Norway to coastal waters amounts to a mean of around 12,000 m³/s for the period 1931–1960 (Tolland, 1976). Figure 4.1 indicates the five main hydrological regions and Table 4.1 gives the normal natural monthly runoff in m³/s for these: Eastern Norway (Region I), Southwestern Norway

Driving forces 47

▶ Table 4.2.
Precentage relative change in the natural runoff due to freshwater regulation by regions (Figure 4.1), 1931–1960 (Modified after Asvall, 1976).

Region	Jan	Feb	Mar	Apr	May	Jun	Jul	Aug	Sep	Oct	Nov	Dec
I	100	170	123	10	-48	-22	-12	5	6	4	14	35
II	26	33	31	12	-11	-19	-12	2	-1	-2	1	12
III	18	14	10	7	-7	-11	-6	-2	-2	-3	8	11
IV	28	38	30	23	-13	-12	-5	-3	-3	-3	10	14
V	24	46	40	29	-2	-12	-14	-3	-2	2	12	15
Total	31	29	32	13	-16	-14	-9	0	0	0	7	15

(Region II), Central Norway (Region III) and Northern Norway (Regions IV and V). The general seasonal variations show low runoff during winter and a marked maximum in May–June. The regions of western and central Norway are characterised by a distinctly rainy autumn season, often giving rise to floods.

Hydroelectric power production may influence the seasonal signal in the freshwater runoff. The variability of the natural freshwater discharge may be more than ± 40 % in summer and ± 20 % in winter. Freshwater regulation, however, will always cause an increase in the winter discharge and a decrease in the summer discharge, which tends to even out the natural seasonal fluctuations. Table 4.2 shows the percentage relative changes in the mean monthly runoff due to freshwater regulation in 1931–1960. As can be seen, this effect is most pronounced in Eastern Norway, where the natural winter discharge may rise by as much as 170 % (Asvall, 1976). In the other regions this effect is significantly less. At local level, however, such regulations may have significant effects on physical and ecological conditions in the fjords.

Climate variations will obviously influence precipitation and thereby freshwater runoff. The main features of long-term variations in the annual runoff include rising values after around 1970 in Regions I to III, while it has remained stable in Regions IV and V. During the same period the winter runoff has either been stable or has shown a slight decrease in most regions, while the spring runoff has increased in all regions. The summer runoff has declined markedly in Regions I and III, while the autumn runoff has increased significantly in Western Norway (Region II) (Førland et al., 2000).

Model simulations predict that for the coming 50 years, as a consequence of expected global climate changes, precipitation in Norway will increase in all seasons, particularly in the autumn (17 %) while in Western Norway the increase is estimated at 23 % (Førland and Nordeng, 1999).

4.2 The tide

Gravitational forces emanating from the moon and the sun cause the tidal elevation of the ocean. The tides observed on the Norwegian coast originate in the Atlantic Ocean and propagate as a tidal wave into the Norwegian Sea and northwards along the coast. Part of the wave turns southwards into the North Sea. In the southern North Sea this wave is reflected and turns north again. In some places, interference between the incoming and the reflected wave creates zones or points with no tidal elevation; so-called amphidromic points. One such point is found off Egersund on the southwestern coast.

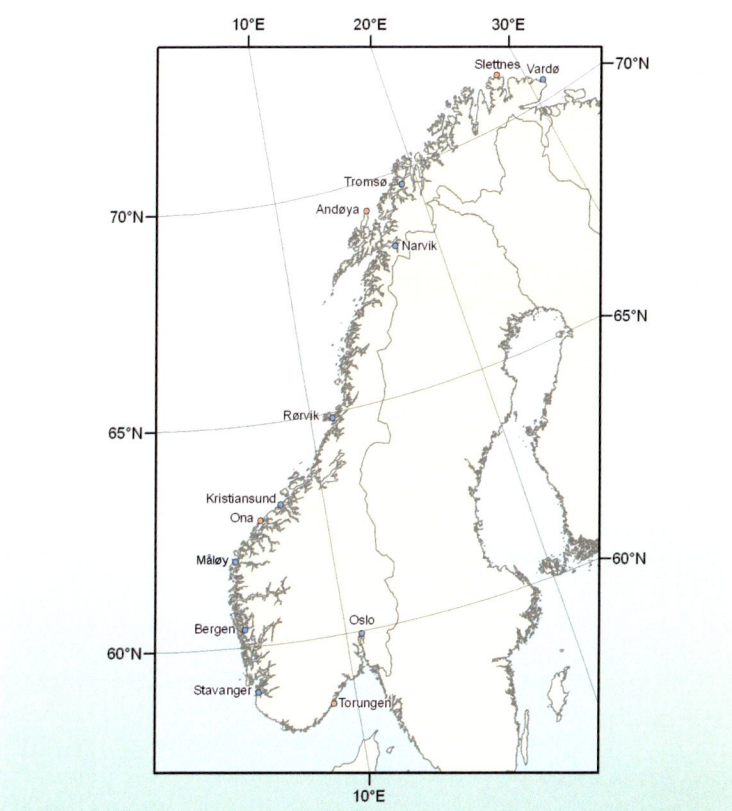

▶ **Figure 4.2**
Locations of meteorological stations and tidal gauges referred to in this chapter.

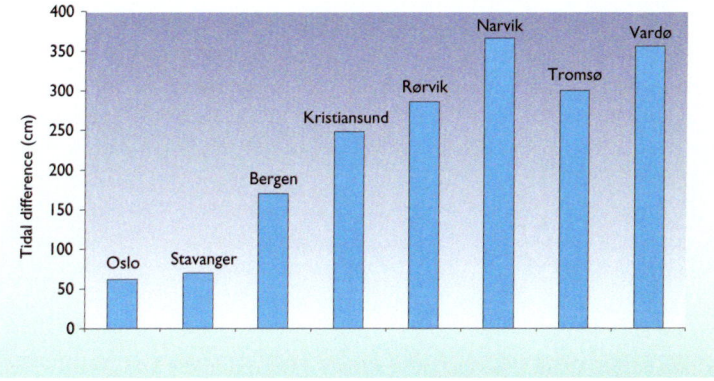

▶ **Figure 4.3**
Mean sea level difference between high and low tides on the Norwegian coast.

The temporal variation in the sea level is made up of a number of harmonic components. On the coast of Norway, the semidiurnal components M_2 (the main lunar component) and S_2 (the main solar component) dominate. Figure 4.3 shows the long-term mean sea level difference between high and low tide on the Norwegian coast as observed from tidal gauges (Figure 4.2). The tidal difference is very small off the southern Norwegian coast. From the Bergen area and northward it gradually increases to reach a maximum of around 3.6 m at Vardø. Due to shallow sills and narrow openings the tidal difference may be greater outside than inside a fjord and the times of high and low tides inside the fjords may be delayed.

Driving forces 49

The times of high and low tides vary along the coast. If Bergen is taken as the reference station, high/low tide will occur approximately six hours ahead of Bergen along the whole of the Norwegian Skagerrak coast. Between Bergen and Måløy the tidal times are about the same, while further north the tide is delayed compared to Bergen, due to the northward propagation of the tidal wave. North of Tromsø the retardation increases, since the tidal wave propagates into the shallower sea area of the Barents Sea, and propagation speed is proportional to depth.

In addition to the gravitational forces of the moon and the sun, meteorological factors such as wind and air pressure may result in variations in sea level. Low air pressures produce high sea levels, while high air pressures decrease them. Persistent southerly and south-westerly winds pile up water against the coast and thus produce a higher sea level. Northerly and north-easterly winds tend to have the opposite effect. As a result, the mean winter sea level is higher than in the summer. In regions where the tidal influence is small, such as off the southern coast and in the Oslofjord, meteorological conditions largely determine variations in sea level. The northerly increase in the tidal difference means that the relative importance of the meteorological conditions decreases northwards along the coast.

4.3 Wind conditions

The most conspicuous climatic feature of the North Atlantic is the North Atlantic Oscillation (NAO), which is expressed in terms of the monthly pressure difference at sea level between the Azores and Iceland (e.g. Blindheim, 2004). The NAO index for the winter months correlates well with both physical and biological conditions in the North Atlantic and has been widely used to study the relationship between physical and biological processes. A high positive NAO index means a strong air pressure gradient between the Azores High and the Iceland Low. This drives strong southwesterly winds, producing mild, wet conditions in northwestern Europe and dry conditions in the Mediterranean. When the negative NAO index is low, the southwesterly winds are weaker, leading to dryer conditions in northwestern Europe and wetter conditions in the Mediterranean.

Norway is situated at the northern margin of the northern westerly wind belt, which means that the winds at sea level along the coast are usually westerly or southwesterly. The "daily weather" is usually a result of disturbances in this westerly airflow. The polar front separates the air masses from the Arctic from those of tropical or subtropical origin. Most of the atmospheric activity, which influences the weather in our areas, take place along this front. Atmospheric low-pressure areas develop along the polar front in the North Atlantic, and these mostly move eastwards or northeastwards towards the Norwegian coast. A special kind of atmospheric lows are what are known as the "Polar Lows", which occur off the coast of northern Norway. These arise under conditions of low air temperature over the open sea. They are generally small but may be very intense between October and April (Iden, 1997).

The coast itself is a boundary zone since both terrestrial topography and temperature differences between sea and land exert a powerful influence on both the direction and strength of the wind. During the summer the temperature over land is higher than over the sea. This generates a wind towards the coast, which is deflected to the right by the rotation of the earth, so that it blows with the land on its left. During the winter, when the land is colder than the sea, the airflow is in the opposite direction. Due to the deflection caused by the earth's rotation it blows with the land on its right side. This tendency to have onshore winds with the land on their left during the summer and offshore winds with the land on their right during the winter is often referred to as the "monsoon effect". When this acts together with the general westerly airflow it may result in an increase or a decrease of the wind speed depending on the season and the general direction of the coastline. Off southern Norway in the summer, the "monsoon effect" tends to augment wind speeds, while it has the opposite effect off the northernmost part (Iden, 1997).

BOX 4.1 THE CORIOLIS FORCE – AN ARTEFACT OF THE ROTATION OF THE EARTH

For an earthbound observer every object moving freely above the surface of the Earth appears to curve slightly from its initial path. The cause of this apparent deflection is called the "Coriolis force" and is a result of the rotation of the Earth. On the northern hemisphere a moving object is deflected to the right while it is deflected to the left on the southern hemisphere (Figure A).

All points on the Earth have the same rotational velocity, as they all rotate once a day. However, places at different latitudes have different linear speeds due to varying distance from the Earth's rotational axis (Red arrow in Figure B). A point near the Equator may rotate at a speed of a thousand kilometres in an hour, while one near the North Pole would move only a few dozen kilometres during the same period of time.

The Coriolis effect is most apparent in the path of a north/south movement. The Earth is rotating from the west to the east. When an object moves north or south it will maintain its initial eastward speed as it moves (Figure B). Objects launched to the north of the equator retain the eastward component of velocity of other objects sitting at the equator. The result is that an object travelling away from the equator will eventually be heading east faster than the ground below (yellow arrow in Figure B) and will apparently be deflected to the east, while an object travelling towards the equator will be going more slowly than the ground beneath it and will appear to be forced westwards (green arrow in Figure A).

The horizontal component of the Coriolis force, C_f, is given by $C_f = v(2\omega \sin\varphi)$, where v is the speed of the moving object, ω is the angular velocity of the Earth's rotation and φ is the latitude. As a result, the strength of the Coriolis force varies according to latitude – it is zero along the equator and increases with increasing latitude and with the speed of the moving object. The Coriolis force affects all movements in the atmosphere and in the ocean. Air moves from areas of high pressure towards areas of low pressure. In the northern hemisphere, such motion is deflected to the right and the result is a vortex of air spinning anticlockwise or cyclonically (Figure C).

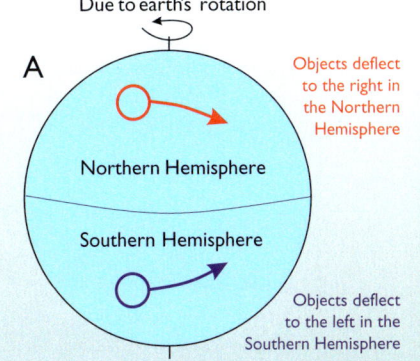

▶ Effect of the Coriolis force on the northern and southern hemispheres.

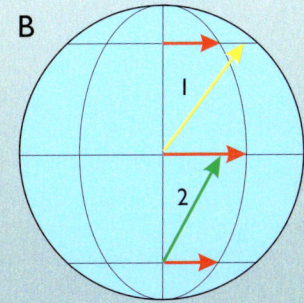

▶ Diagrammatic outline of how the Coriolis force influences a northwards movement.

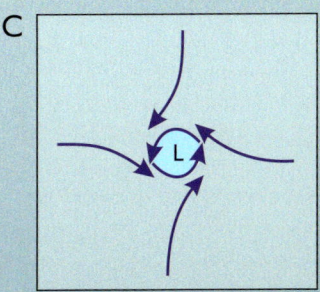

▶ The anticlockwise spinning of the air around an atmospheric low-pressure area is another effect of the Coriolis force.

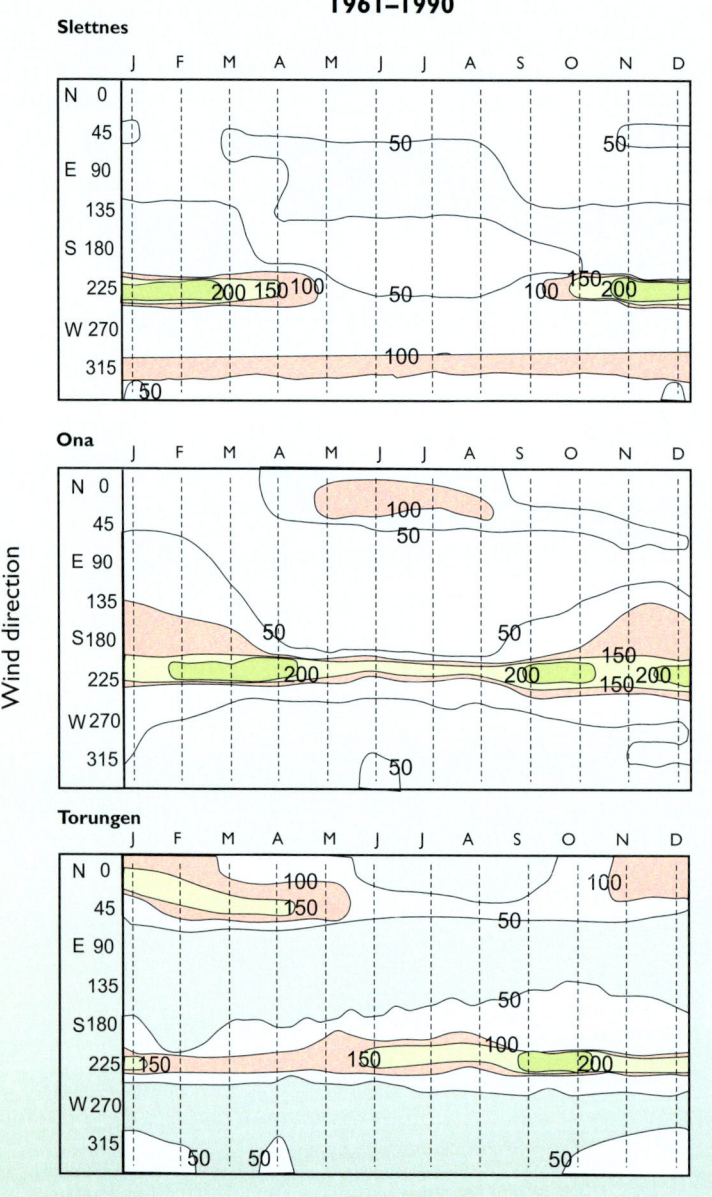

► Figure 4.4
Monthly mean of wind amount (Frequency x mean wind force) at selected meteorological stations during the mean year 1961–1990.

In addition to these more general effects on the wind field, local topography may have pronounced effects on both the strength and direction of the wind. Around headlands, especially if they are high with steep sides, the wind speed will increase due to the "corner effect". Local increases in wind speed are quite common in areas where fjords or sounds become narrower.

Winter is characterised by winds that usually blow with the land on their right when we look in the wind direction and the wind is blowing seawards in the fjords. During the summer winds with land on their left are the most frequent and these are blowing into the fjords. Further details are seen in Figure 4.4, which shows monthly means of amount of wind in different directions (frequency or duration x mean wind force in Beaufort) for the meteorological years 1961–1990 at three locations (Figure 4.2) on the coast. The station Slettnes represents wind conditions on the Barents Sea coast and Ona, winds on the coasts of the North Sea and Norwegian Seas, while Torungen is representative of the Skagerrak coast. They all show the seasonal "monsoon" variability pattern described above, in which two main wind directions dominate.

All along the coast, except the Skagerrak, northeasterly winds dominate during the summer months, while southwesterly winds are more frequent and have higher force during the winter. At Torungen on the Skagerrak coast, northeasterly winds dominate during the winter while southwesterly winds are more pronounced in the summer (Figure 4.5). A characteristic feature at Torungen is the strong summer winds, which are caused by the heating of the land during the daytime and thus a "monsoon effect" on a daily basis.

During the 1990s the NAO index has been very high, which naturally raises the question of whether there has also been a shift in the wind pattern in the 1990s along the coast in comparison with the previous decades. Figure 4.6 shows the difference in amounts of wind for the two main wind directions between the annual mean for 1991–2000 and the mean for 1961–1990 for the meteorological stations shown in Figure 4.2. At Torungen there has been a drastic decrease in the amount of northeasterly winds and a similar increase in southwesterly winds during the winter for the 1991–2000 period, compared to 1961–1990. For the rest of the year there are only minor changes except in September, the last decade has seen a large increase in northeasterly winds.

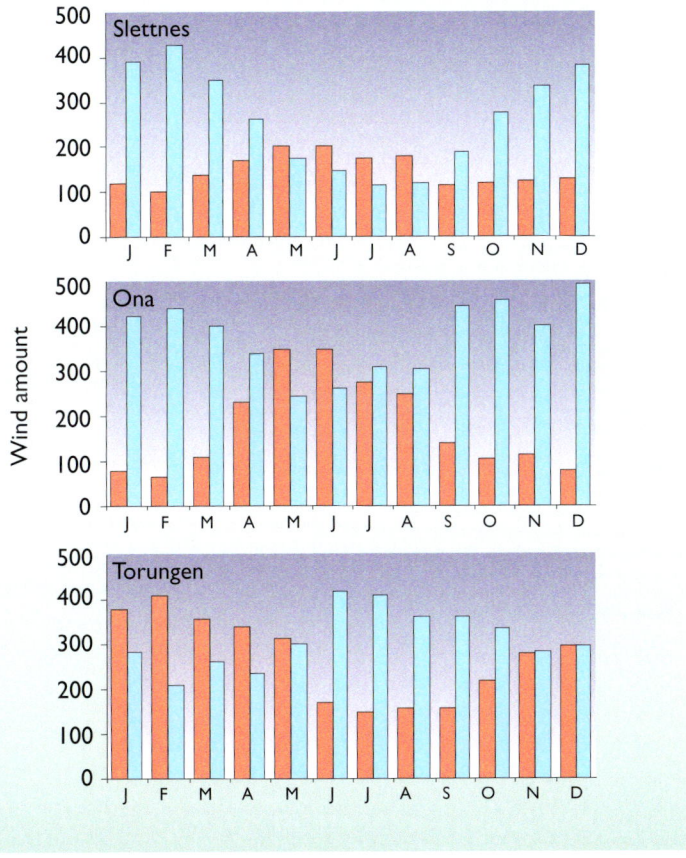

▶ **Figure 4.5**
Seasonal variations in the wind amount along the two main directions on selected meteorological stations during the mean year 1961–1990.
Red columns: Northeasterly wind amount.
Blue columns: South or southwesterly wind amount.

Driving forces 53

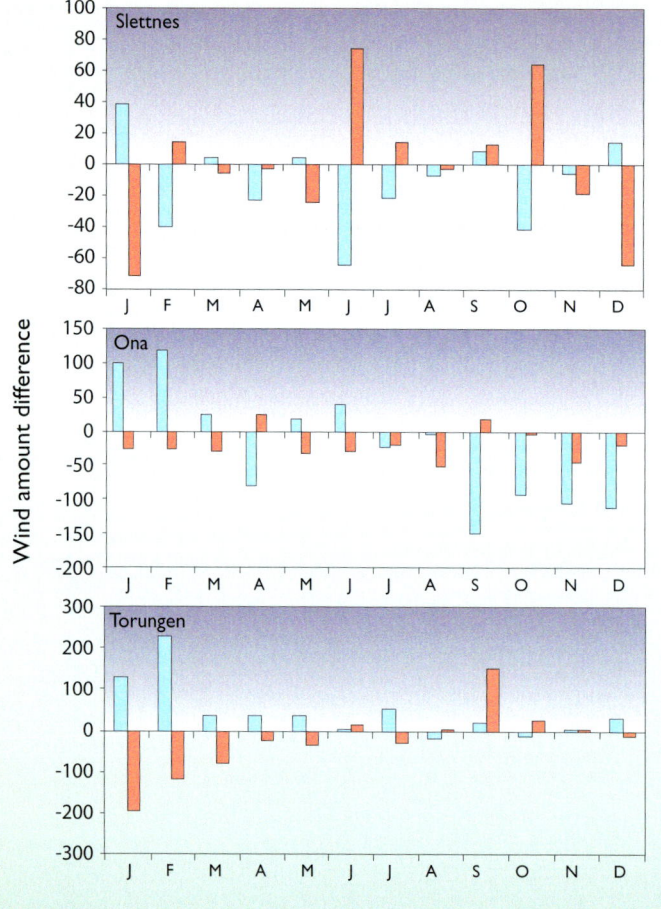

► **Figure 4.6**
Difference in wind amount for the two main wind directions between the mean years 1991–2000 and 1961–1990 for selected meteorological stations.
Red columns: Northeasterly wind amount.
Blue columns: South or southwesterly wind amount.

Ona shows a significant increase in the mean amount of southwesterly wind in 1991–2000 during the first three months of the year, while there is a similar decrease in the autumn. In the northeasterly wind amount there are only minor changes and these mostly show a small decrease. Slettnes shows an increase in northeasterly winds during June and October and a decrease in December and February. In most months there is a decrease in the amount of southwesterly wind. The relative difference in the amounts of wind for the two periods are highest for Torungen and Ona, while it is rather small for Slettnes.

4.4 The Atlantic influence

The relatively warm and saline Atlantic water flows into the Norwegian Sea from the North Atlantic Ocean mainly through the Faeroe-Shetland Channel (Figure 4.7). Another branch of the Atlantic inflow crosses the Greenland-Scotland Ridge north of the Faeroe Islands. These water flows transport a tremendous amount of heat into and through the Norwegian Sea. This has been estimated at around 250 TeraWatts, equivalent to 15 to 20 million times the mean production of electric energy in Norway (Blindheim, 2004). A significant part of this energy flow is released in the Nordic Seas, producing the very mild climate of north-western Europe compared to the same latitudes on the western North Atlantic Ocean.

The Atlantic inflow to the Nordic Seas is mainly forced by the atmosphere (wind and air pressure). There is also some deep convection in areas inside the Nordic Seas, where surface cooling forms deep or bottom water. These water masses leave the Nordic Seas across the Scotland–Greenland Ridge, providing an extra forcing of Atlantic water inflow to compensate this outflow.

The circulation of the inflowing Atlantic Water is greatly influenced by the bottom topography. It continues northward as the Norwegian Atlantic Current, with its main flow along the edge of the Norwegian continental slope. Its mean current speed is 30–35 cm/s while its maximum current speed quite frequently

BOX 4.2 SOME BASIC DYNAMIC OCEANOGRAPHIC TERMS

The **hydrostatic pressure**, p, at any depth below the sea surface is given by $p = g\rho z$, where g is the acceleration of gravity, ρ is the density of seawater, which increases with depth, and z is the depth below the surface. Horizontal differences in density due to variations in temperature and salinity cause the hydrostatic pressure to vary in the horizontal plane and thereby create horizontal pressure gradients. Although these are much smaller than the vertical changes in pressure, they give rise to ocean currents.

In an ocean with no horizontal density differences horizontal pressure differences are possible only if the sea surface is tilted. In such a case, the surfaces of equal pressure are tilted in the deeper layers by the same amounts as the sea surface. This is referred to as the **barotrophic field of mass**. The horizontal pressure gradient due to the sloping sea surface produces a current speed that is independent of depth (Figure A).

Horizontal variations in temperatures and salinity cause the horizontal pressure gradient to vary with depth. This is the **baroclinic field of mass**, which leads to currents that vary with depths (Figure B). The horizontal pressure gradients in the ocean are a combination of these two mass fields.

For most of the ocean volume away from the boundary layers frictional forces are of minor importance, and the equation of motion for horizontal forces can be expressed in terms of a simple balance of the horizontal pressure gradient and the Coriolis force. On a rotating Earth the Coriolis force, C, deflects the motion, and the acceleration ceases only when the speed, v, of the current is just fast enough to produce a Coriolis force, C, that exactly balances the horizontal pressure gradient dp/dn (Figure A). This type of current is called **geostrophic current**, and from this balance it follows that the current direction must be perpendicular to the pressure gradient because the Coriolis force always acts perpendicularly to the direction of motion. The geostrophic current represents a state of least energy on a rotating Earth. This is the "the preferred state" or the "path of least resistance".

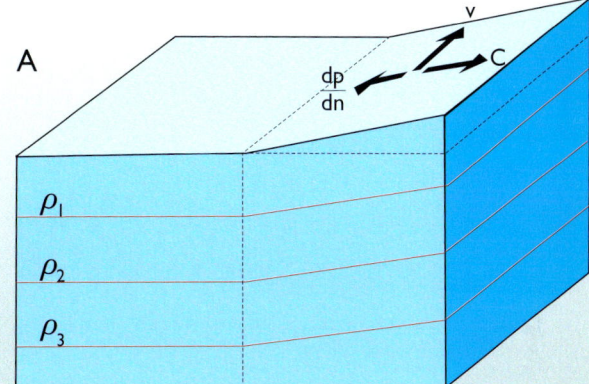

▶ A barotrophic field of mass distribution.

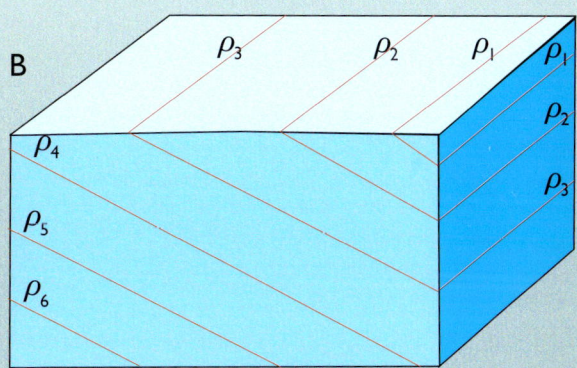

▶ A baroclinic field of mass distribution

Driving forces

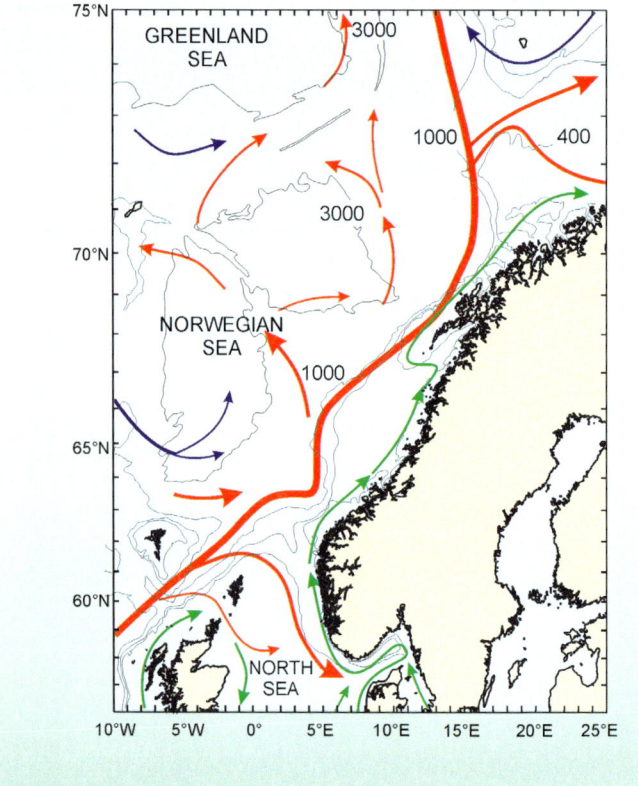

▶ **Figure 4.7**
The main circulation pattern of the Nordic Seas.
Red arrows:
Atlantic water.
Blue arrows:
Arctic water.
Green arrows:
Coastal water.

exceeds 100 cm/s. The mean annual transport has been calculated at 4.2 Sverdrup (4.2 10^6 m^3/s) (Orvik *et al.*, 2001).

The Norwegian Atlantic Current flows northwards, parallel to the wedge-shaped Norwegian Coastal Current and Atlantic water covers the area below and on the seaward side of the Norwegian Coastal Current (Figure 4.8). Along the route there is dynamic interaction, such as eddies, between the two currents. There is also a transport of heat, salt and nutrients from the Atlantic Water into the Coastal Water.

There are seasonal and inter-annual variations in both the properties and the volume transport of the Atlantic Water. The volume transport during the winter may be twice as high as during the summer. The interannual fluctuations may be due to both regional effects such as wind conditions or cooling and variations in the properties of the inflowing Atlantic Water. Such long-term, climatic fluctuations are of great biological importance for both the ecosystem of the Norwegian Sea and that of the Norwegian coast.

4.5 Bottom topography

Bottom topography is important in relation to circulation and vertical mixing. Water flows have a tendency to follow the depth contours with the shallowest part on the right when facing the direction of the current. Figure 4.9 shows the principal depth conditions off the coast of Norway. Off the southern part of the country the dominant bathymetric feature is the Norwegian Trench, leading from the Norwegian Sea at approximately 62°N into the Skagerrak. At its northern

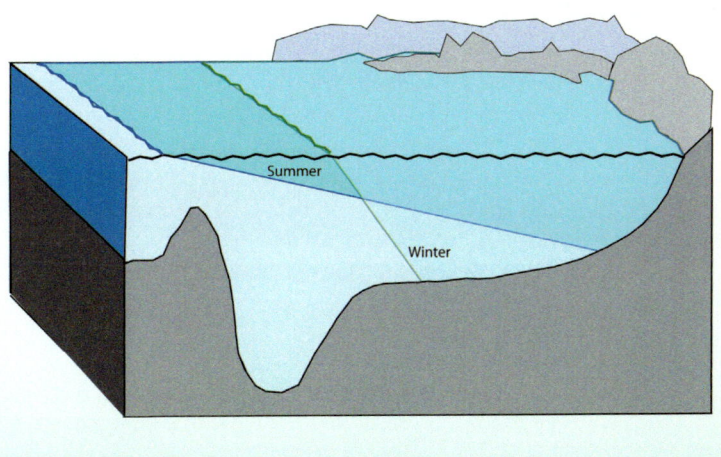

▶ **Figure 4.8**
The coastal wedge – summer and winter.

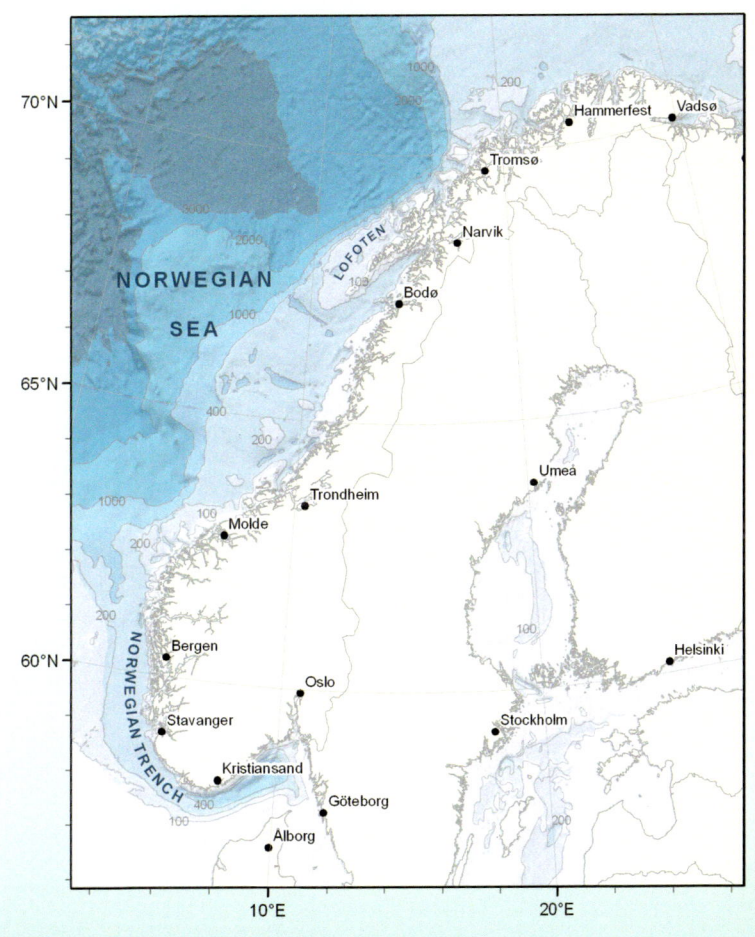

▶ **Figure 4.9**
Bathymetry of the Norwegian continental shelf.

Driving forces 57

end depth range between 400 and 500 m, while in the Skagerrak depths can be greater than 700 m. The Norwegian Trench has a sill of about 270 m around 59°N. To the west and the south there is a gradual slope up to the shallower North Sea plateau of only around 50 m.

Between 62° and 68°N the Norwegian continental shelf is at its widest, with an average width of more than 200 km. The area out to the 500 m depth contour covers about 150,000 km^2 or about 25 % of the total area of the North Sea. The bottom topography is complicated, consisting of several banks that are often less than 100 m deep, separated by deeper channels of more than 200 m. There are also large depressions and troughs more than 400 m deep that are aligned transverse to the coast. Off the Lofoten archipelago the continental shelf narrows to a minimum width of about 2 km off Andøya. Further north it widens again to include a number of shallow banks separated by deeper channels. Off the northernmost coast bordering the Barents Sea the bottom is more regular again, with depths of 200 to 400 m.

The fjords of Norway are generally deep, with a shallow sill at the entrance. The longest of them, the Sognefjord, is also the deepest, reaching a depth of more than 1300 m and with a sill of about 200 m. In northern Norway some of the large fjords are without sills. They thus lack the typical fjord circulation and should therefore be regarded rather as oceanic bays.

Properties of coastal water masses

Roald Sætre

A water mass is characterised by a temperature-salinity relationship. When temperature observations from a specific area are plotted against the associated salinities, the points will lie along a curve that characterises the water properties of that location. Along the Norwegian coast two main water masses dominate; the Atlantic Water and the Norwegian Coastal Water. According to a generally accepted definition, water of salinity greater than 35 is called Atlantic Water and that of salinity below 35, Coastal Water. However, along most of the Norwegian coast a salinity of 34.5 represents a better separation between Coastal and Atlantic waters. Both the temperature, salinity and the hydrographic conditions of the Coastal Water vary in space and time. In addition to fluctuations due to geographical location the temporal variation may be split into short-time variations (time-scale of up to approximately one month), seasonal variations and long-term variations (over several years). In this chapter we deal with the mean geographical and seasonal fluctuations.

5.1 The upper layers

The temperature and salinity observations in the surface layer along the Norwegian coast, known as the Thermograph Service, started in 1936 from coastal steamers. The historical background of this observation system is described in Chapter 2. Figure 5.1 shows the main observation points. The positions north of Bergen are still being operated, while most of the observations along the southern Norwegian coast were terminated in the late 1970s due to the reduction in the number of coastal steamers. Midtun (1971), Sætre (1973) and Aure and Østensen (1993) have present aggregated data from the Thermograph Service.

Figure 5.2 illustrates the mean geographical variations in seasonal extreme values of temperature and salinity in the surface layer along the coast. In order to have a homogeneous dataset covering the whole coast, the mean year 1936–1970 was chosen (Sætre, 1973), as this is believed to provide a good characterisation of general hydrographic conditions in the surface layer. The maximum temperature usually occurs between 25 July and 5 September and the minimum between 15 February and 5 April. Maximum salinity is observed during the winter, from December to April, and minimum salinity between May and October. This pattern is related to the freshwater outflow to the coastal current. At locations with good water exchange with the open ocean, the maximum and minimum values usually occur later than at locations where water exchange is more restricted. There is also a tendency to a northward delay in the timing of seasonal extreme values.

As can be seen, the seasonal maximum temperature falls as we move northwards. The lowest minimum temperature, or winter temperature, is observed on

▶ Figure 5.1
Locations of the main stations for observations of surface temperature and salinity on the Norwegian coast.

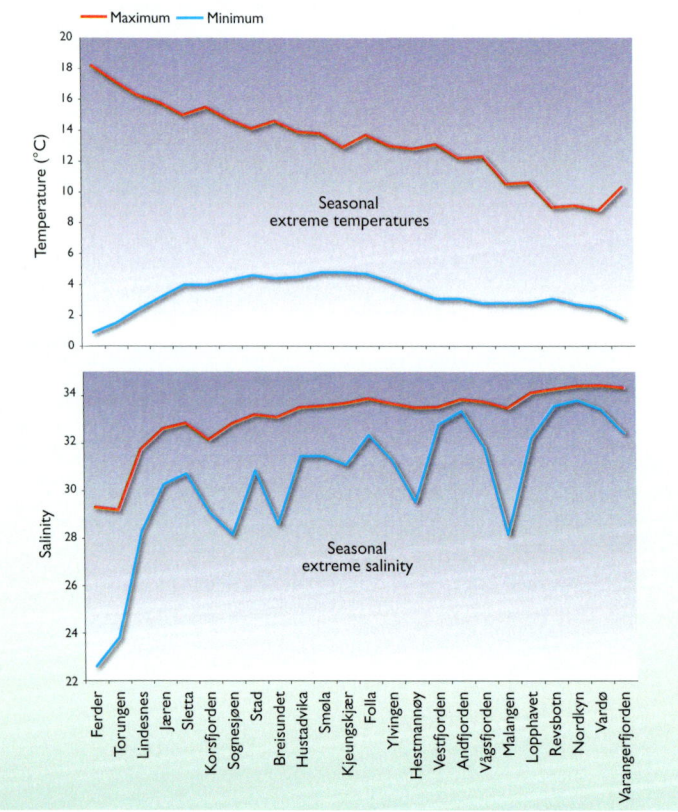

▶ Figure 5.2
Mean seasonal extreme values of temperature and salinity on the Norwegian coast based on the observation points of Figure 5.1.

the Skagerrak coast, and the highest winter temperatures are observed off central Norway. This is clearly due to the strong Atlantic influence in this region. Both maximum and minimum salinities increase as we move north, with the steepest geographical increase from the Skagerrak to Jæren. This is due to large amounts of fresh water from the Baltic (Chapter 4) being rapidly entrained with salt water within the Skagerrak. For the salinity minimum, however, there are wide local variations. This minimum is related to the large freshwater runoff to the fjords during the summer. Observation points that are strongly influenced by the fjords will therefore have significantly lower salinity values during the summer than observations points with good connections to the open ocean. On the Norwegian Skagerrak coast, ice formation in the coastal area may be a problem in severe winters, particularly in the eastern part.

Further details can be seen in Figures 5.3 and 5.4, which show the mean seasonal variations in the upper layer temperature and salinity for the whole coast, using data from the observation points of Figure 5.1. The major increase in the salinity of the coastal water after it has left the Skagerrak and the low salinities at the entrances of some fjords are conspicuous. The mean seasonal temperature curves along the coast appear to be asymmetric, as the time of cooling is longer than the time of heating. This feature is related to the vertical layering of the water masses and thereby to vertical mixing and mixing agents such as wind speed. In northern Norway the cooling time in the autumn is approximately twice as long as the heating time during the spring.

The typically wedge-shaped Norwegian Coastal Current (Figure 4.8) shows a clear seasonal profile, where the wedge is deep and narrow during the winter and

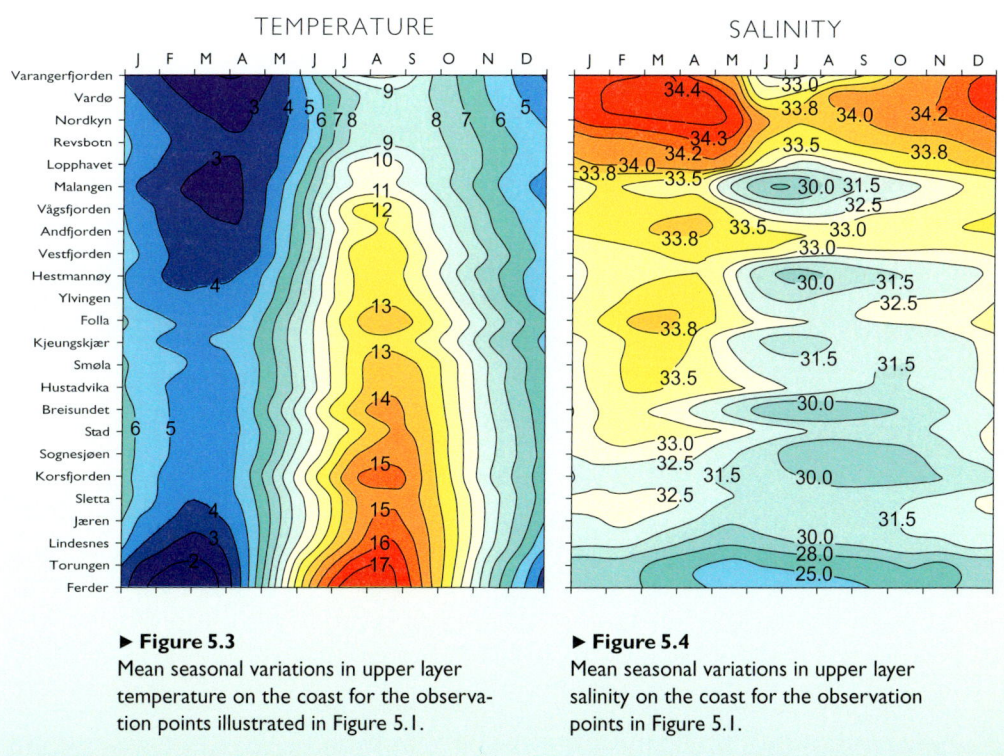

▶ **Figure 5.3**
Mean seasonal variations in upper layer temperature on the coast for the observation points illustrated in Figure 5.1.

▶ **Figure 5.4**
Mean seasonal variations in upper layer salinity on the coast for the observation points in Figure 5.1.

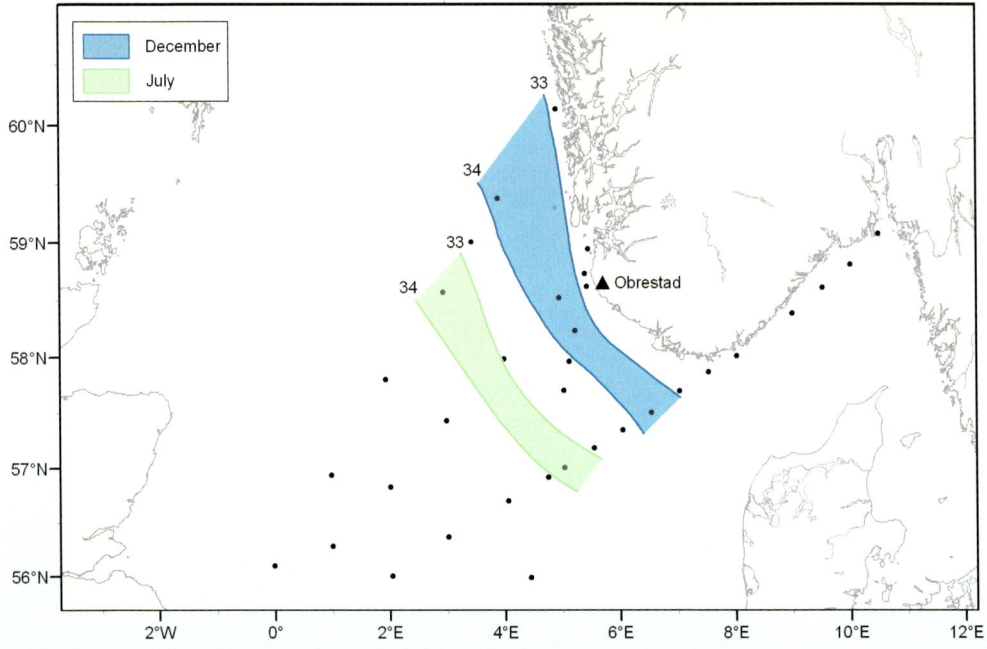

▶ **Figure 5.5**
Mean positions of the 33 and 34 isohalines in the surface layer for December and July based on ships of opportunity observations (Figure 2.7).

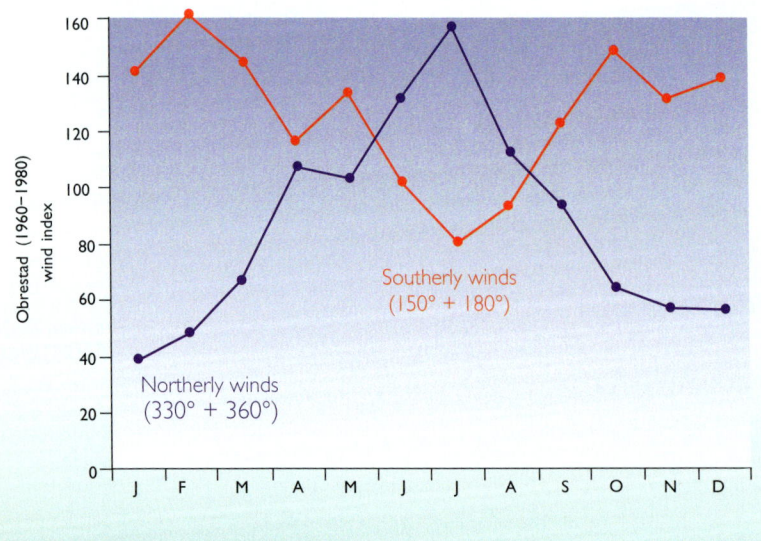

▶ **Figure 5.6**
Mean seasonal variation in the northerly and southerly wind index (Frequency or duration x average wind force in Beaufort) at Obrestad.

wide and shallow in the summer. This means that the coastal water at the surface moves farther from the coast during the summer than the winter. The long-term mean December and July positions of the 33 and 34 isohalines off southwestern Norway (Figure 5.5) demonstrate the seasonal horizontal displacement of the Norwegian coastal water in the surface layer. The mean seasonal horizontal displacement of the coastal surface layer off southwestern Norway is 80–100 km (Sætre et al., 1988). Figure 5.6 demonstrates the pronounced seasonal variations in the mean northerly and southerly wind amount or index (direction frequency x average wind force) at Obrestad (Figure 5.5). The northerly wind index in July is approximately three times as high as that of December.

Several reasons for the seasonal displacement of the Norwegian Coastal Current, which was first observed by Hjort and Gran (1899), have been suggested in the past. These include lower summer densities, due to heating and increased fresh water influx, seasonal variations in the vertical eddy diffusion and the prevailing winds. These potential generating forces are not necessarily independent. Sætre et al. (1988), however, concluded that the seasonal western displacement of the coastal water is probably due to Ekman transport by the monsoon-like wind pattern on the Norwegian coast, with increased northerly wind stress during summer. The "monsoon effect" is less pronounced off the northernmost parts of Norway, and thus also the lateral seasonal displacement of the coastal water.

The minimum salinity on the coast of southern Norway occurs in May–June, coinciding with the time of maximum freshwater discharge. Along the western coast north of Stavanger, however, the minimum surface salinity close to the coast occurs in September–October, while in the Norwegian Trench and to the west it occurs in June. This indicates a westward displacement of the Norwegian Coastal Current during the summer, with coastal upwelling (Sætre et al., 1988).

The most important factor that determines average seasonal variations in salinity off southern Norway appears to be variations in the Baltic outflow. The annual freshwater discharge from the southern Norwegian coast north to Utsira, during a mean year, is about 23 % of the annual fresh water outflow from the Baltic (Sætre, 1979). On a mean monthly basis the freshwater runoff south of Utsira is between 10 and 15 % of the freshwater outflow from the Baltic from January to August and between 20 and 40 % from September to December. The variability in the Baltic outflow is gradually masked as we move north along the coast of Norway, and is difficult to trace north of about 62°N.

At some of the observation points along this part of the coast, in the Skagerrak/Kattegat area and at some of the observation points along the shipping routes across the North Sea (Figure 2.6), a secondary salinity minimum is observed, usually in August. Figure 5.7 shows the area of occurrence of the secondary salinity minimum (Sætre, 1979), which indicates its connection to the Baltic outflow. The outflow of coastal water from the Skagerrak is intermittent, reflecting similar fluctuations in the Baltic outflow. There is a maximum in the frequency of sudden occurrences of Baltic water on the Norwegian Skagerrak coast during February and August, indicating similar maxima in the Baltic inflow to the Skagerrak (Aure and Sætre, 1981). Ljøen (1981) found that the extension of water of salinities below 34 in the Torungen–Hirtshals section across the Skagerrak has two maxima; one in February and one in September, which he also related to similar maxima in the Baltic outflow. Figure 5.7 also shows that the outflow from the Skagerrak occasionally has a tendency to follow a more southerly and westerly route during the summer, as is also shown by the mean surface salinity maps for the summer presented by Ljøen (1988).

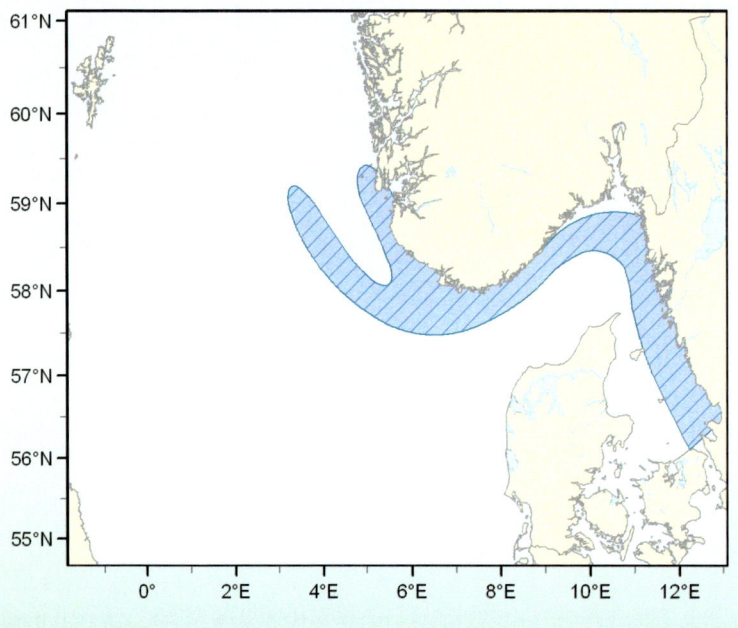

► **Figure 5.7**
Area of occurrence of the secondary salinity minimum during the mean year 1967–1976.

According to Sætre (1979), the increased westerly wind stress in the Skagerrak in June and July will retard the Baltic outflow. This wind situation is also related to low atmospheric pressure over the Baltic, which increases its water-storage capacity. When the westerly wind stress is reduced in August, the outflow increases, resulting in a secondary salinity minimum. The apparent splitting of the area for the secondary salinity minimum between Kristiansand and Stavanger is probably a combined effect of the topography and peculiarities in the wind field in this area. This stretch of the coast appears to be the location of the most regular and persistent coastal upwelling anywhere on the entire Norwegian coast (e.g. Eggvin, 1940).

5.2 The lower layers

The Norwegian coastal observation system consists of a number of fixed hydrographic stations (Figure 2.9) operated by local observers, and several fixed hydrographic transects running approximately perpendicular to the coast. Figure 5.8 shows the location of those that were used to produce the following description. Although the number of fixed oceanographic stations and sections shown is low, they adequately illustrate the regional differences.

The Skagerrak coast

Most of the water entering the North Sea from outside leaves it via the Skagerrak. Hydrographic events that take place in the North Sea are thus reflected in the Skagerrak. The currents in the region form a large anticlockwise circulation (Figure 5.9), in which the water column flows in the same direction from the surface to the bottom. A number of different water masses from the North Sea and the Kattegat enter the Skagerrak and largely mix, making their origin difficult to determine. The water masses of the area are usually divided into three: the

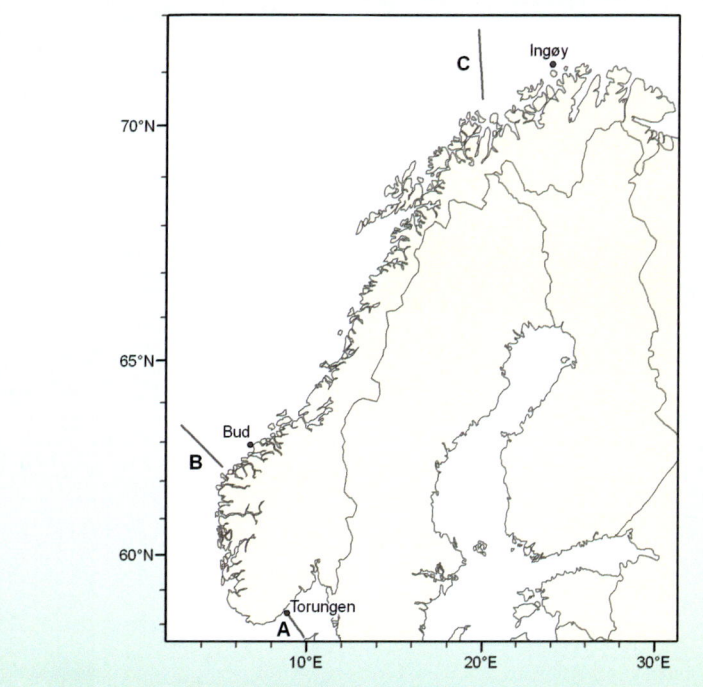

▶ **Figure 5.8**
Location of the fixed hydrographic stations and transects referred to in this chapter.

▶ **Figure 5.9**
The general circulation pattern of the Skagerrak and the Kattegat (Anon, 1993).
AC: Atlantic Current.
NJC: North Jutland Current.
KSC: Kattegat Surface Current.
KDC: Kattegat Deep Current.
SCC: Skagerrak Coastal Current.
Recirc: Resirculation of Skagerrak Coastal Water (SCW) and Skagerrak Water (SW).

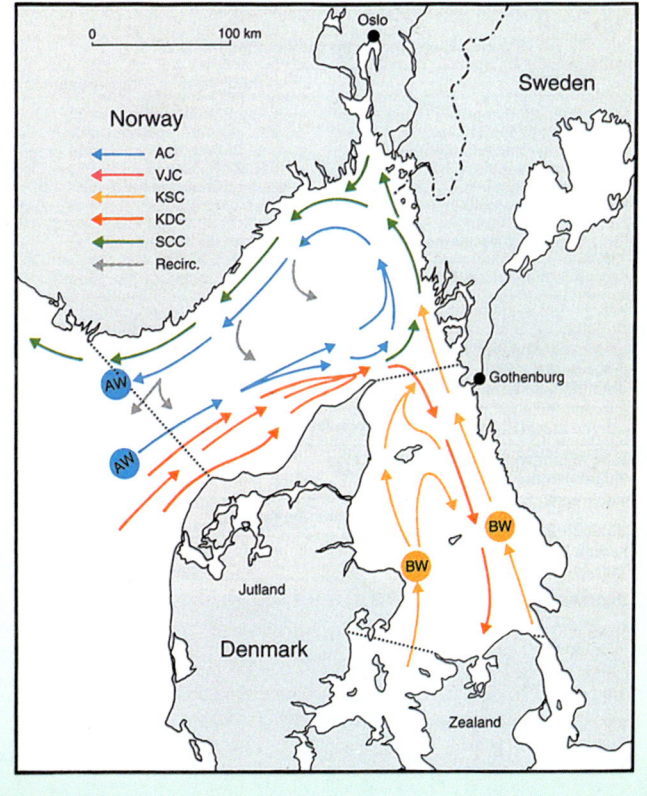

Properties of coastal water masses 65

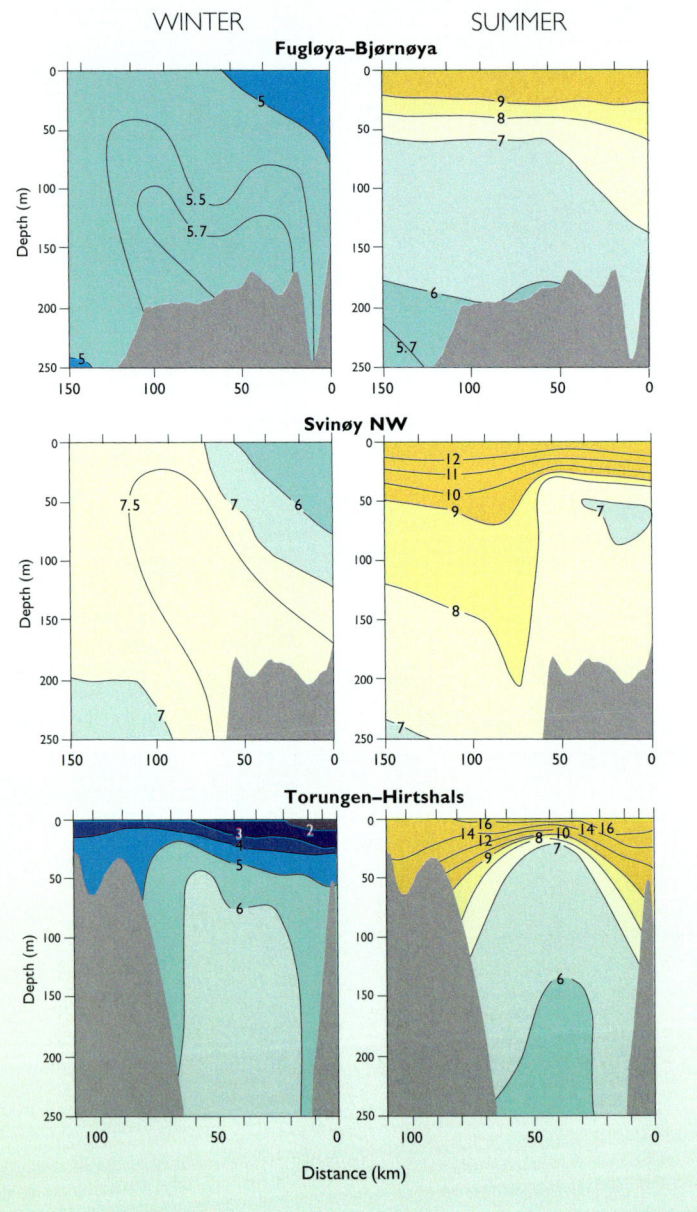

▶ Figure 5.10
Long-term mean winter and summer temperatures at fixed hydrographic sections Fugløya–Bjørnøya (C) Svinøy (B) and Torungen–Hirtshals (A) (Figure 5.8).

Skagerrak Coastal Water (SCW) with salinities of 25–32, which mainly occupies the northern and eastern part, the Skagerrak Water (SW), with salinities around 32–35 and the Atlantic Water (AW), with salinities above 35.

Figures 5.10 and 5.11 show the long-term mean temperatures and salinities of the Torungen–Hirtshals Section for March and August. The surface layers, with salinities of 25–32 are thin, usually less than 20 m deep, and are thicker along the Norwegian coast as a result of the influence of the Baltic water. Below the surface layer there is a core of water with salinities of 32–35, which has its origins in the North Sea. In the deeper layers water of Atlantic origin and of salinity greater than 35 is seen. A characteristic feature of the hydrographic structure of

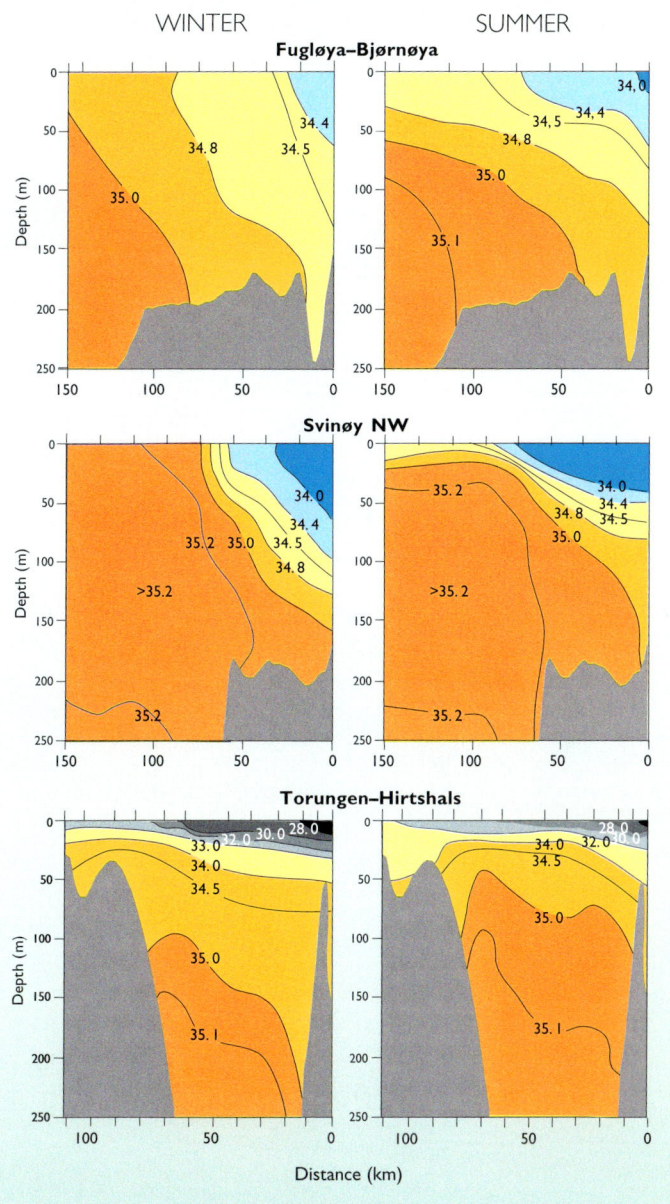

▶ Figure 5.11
Long-term winter and summer mean salinities at the fixed hydrographic sections Fugløya–Bjørnøya (C) Svinøy (B) and Torungen–Hirtshals (A) (Figure 5.8).

the Skagerrak is its domed shape, with higher salinities and lower temperatures in the central part of the Skagerrak, particularly during the summer months. The large and slightly warmer sub-surface volume in the middle of the trench during the winter indicates that this water mass has a much longer residence time than the water masses along the slopes (Danielssen et al., 1996).

It seems that the Atlantic water masses (salinities above 35) reach their maximum extent during the summer. In contrast, water of salinities between 34 and 35, originating from the central and the northern North Sea, appears to be most widely distributed during the summer. The waters with salinities below 34, however, have two maximum extensions; one in February and another in

BOX 5.1 WHY IS THE OCEAN SALTY?

According to a Norse fairy tale it is because somewhere on the bottom of the ocean there is a mill grinding salt. In fact, the story gives a reasonable picture of what is happening – it needs only some extra details.

It was once believed that the saltiness of the ocean is simply due to the storage of salts derived from rock weathering and transported to the ocean by fluvial processes. Though these processes are responsible for nearly all the salts in the ocean, they cannot alone explain the composition of seawater. Important elements such as chlorine, bromine and iodine are missing in these effluents. We therefore need to identify additional sources of ocean salts. Transport of elements from the interior of the Earth through the seabed can explain the composition of seawater.

The Earth's crust is divided into several plates, which move as rigid units at speeds of several cm/year (Figure A). The fracture zones between these plates are associated with earthquake and volcanic zones and they are often related to subsurface mountainous ridges. The 6000 km S-shaped Mid-Atlantic Ridge is such a fracture zone. Along these ridges there is usually a Ridge Valley where the two oceanic plates are being pulled apart and erupting lava replaces the sea floor. The extremely hot mineral-rich fluids from beneath the ocean floor flow through cracks in the crust (Figure B). This flow carries with it "juvenile" water, i.e. water derived from the interior of the Earth, which has not previously existed as atmospheric or surface water. Another type of juvenile water is released by volcanic action. The elements missing from the effluents from land make up most of the dissolved substances in juvenile water.

The composition of seawater has probably not changed for several billion years, due to the equilibrium between input and removal of salt. This means that the ocean is not simply an accumulator of salts, but that, as water evaporates from the ocean, along with some salt, the introduced salts are removed in the form of minerals. Thus, the concept of the ocean as a chemical system has changed from that of a simple accumulator to that of a steady-state system, in which rates of inflow of materials into the ocean equal rates of outflow. The steady-state concept permits influx to vary with time, but it is matched by nearly simultaneous and equal variations in efflux.

This means that there needs to be a balance between the geochemical processes that take part in the salt cycle of the ocean. These processes include exchange between the ocean, the atmosphere, rivers, rocks, sediments on the ocean floor and the plastic interior of the Earth. For example, there is a transport of salt from the ocean to the atmosphere where the salt particles act as condensation cores for growing raindrops. This process is responsible for the main part of the salts transported to the ocean by the rivers. If the hypothesis of an ocean unchanging over time is correct, it means that the introduced salts must form a part of the marine sediments.

On average, the ocean contains 3.5 % salt, of which sodium chloride (NaCl) or ordinary table salt is the dominant component. If all the salt were to be extracted from the ocean today and dried, it would form a 45 m thick layer over the whole planet. At least 72 chemical elements have been identified in seawater, most in extremely small proportions. Probably all of the Earth's naturally occurring elements exist in the sea. Apart from salt and fresh water, only bromine and magnesium are extracted from seawater in significant amounts. Some marine organisms have the ability to accumulate and thereby concentrate substances from the sea. Iodine, for example, was observed in marine algae 14 years before it was identified in seawater. Inorganic processes may also concentrate substances such as the manganese nodules observed at the bottom of most of the deep oceans.

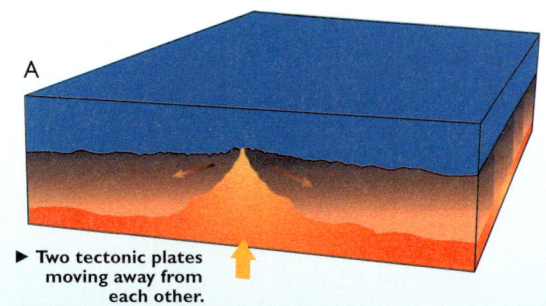

▶ Two tectonic plates moving away from each other.

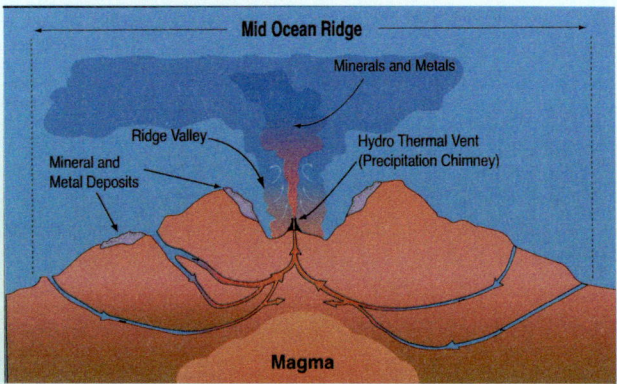

▶ Lava erupting from the seabed at a Mid-Ocean Ridge.

September, which are clearly associated with similar variations in the Baltic outflow (Ljøen, 1981).

The deeper parts of the Skagerrak may remain stagnant for several years at a time, interrupted by short periods of renewal. The renewal of the Skagerrak deep water may be an effect of cooling of the water on the North Sea plateau during extremely cold winters. The water will then be heavy enough to cascade down to the deeper parts. The deep water may also be renewed by inflows of heavier Atlantic Water.

Figure 5.12 shows the mean seasonal variation in temperature and salinity at the fixed oceanographic station of Torungen. The minimum temperature of the surface layer is about 1.4 °C in February–March. By turbulent mixing this

▶ **Figure 5.12**
Long-term mean seasonal variations in temperature and salinity for the fixed oceanographic stations at Ingøy, Bud and Torungen (Figure 5.8).

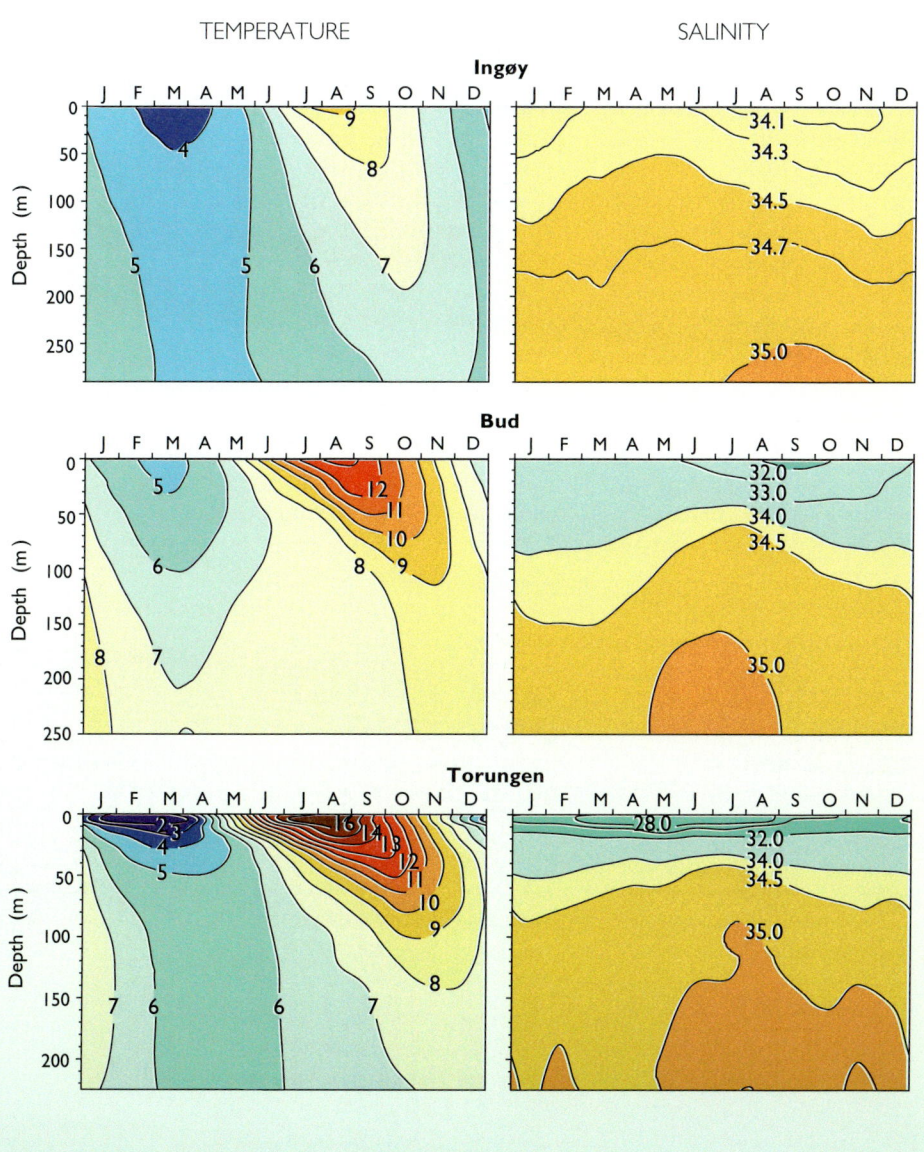

Properties of coastal water masses 69

minimum propagates rather rapidly to the deeper layer, reaching a depth of 200 m in April. The maximum temperature, approximately 17 °C, occurs in the surface layer in August. It propagates much more slowly to the deeper layers and takes four months to reach a depth of 200 m. The reason for this difference in the vertical propagation time for the two seasonal extremes is the difference in the vertical stability of the water masses between summer and winter. The vertical turbulent mixing is much more effective during winter when the vertical stability is low.

The minimum salinity in the upper 20 m occurs in April–May, while at depths of 50 m or more it usually occurs in January. Salinity rises rapidly with depth, and by a depth of 50 m it is higher than 34 all year. The maximum salinity in the upper 20 m is registered in December–January, and in the deeper layers, usually in July. The apparent discontinuity around 30 m depth in the occurrence of seasonal extremes in salinity indicates that the waters of the upper layers, the Skagerrak Coastal Water, to a large degree lives its own life.

The central Norwegian coast

The central Norwegian continental shelf between 62°N and 68°N is rather shallow and has an irregular bottom topography (Figure 4.9). The deeper layers of the shelf are usually covered by Atlantic Water, while the Coastal Water dominates the upper layer. The seasonal temperature and salinity variations at the fixed oceanographic station of Bud (Figure 5.8) are regarded as representative of the hydrographic conditions in the coastal areas of this region (Figure 5.12). The monthly minimum temperature in the surface layer is slightly below 5 °C and occurs in March. This minimum is transported to deeper layers by turbulent mixing and reaches the seafloor after approximately one month.

The mean monthly maximum temperature is 12–13 °C and is usually recorded in August. The vertical transport of the temperature maximum is much slower, taking about four months to reach the bottom.

The minimum salinity of around 32 in the surface layer is found in September and the maximum of about 33.5 in March. Atlantic Water with salinity above 35 is observed below 150 m during the summer. The 34.5 isohaline is regarded as being more representative of the transition layer between the Atlantic and Coastal waters. As Figure 5.12 shows, the depth of this isohaline at Bud varies between 60 m during the summer to 150 m during the winter.

Figures 5.10 and 5.11 show the long-term mean vertical distribution of temperature and salinity during the winter and summer in the Svinøy section (Section B–Figure 5.8). In both seasons, the Atlantic Water core, with salinity above 35.2 and a temperature of 7° to 9 °C over the continental slope, is conspicuous. The lower depth of this core is about 150 m. The Atlantic Water covers the deeper waters over the shelf in both winter and summer. As can be seen in the figures, the Coastal Water has a more seawards distribution during the summer than the winter and consequently the Coastal Water wedge is narrow and deep during the winter while it is wide and shallow in the summer (Figure 4.8). This feature is also reflected in the salinity distribution at Bud, where the Atlantic Water rises during the summer (Figure 5.12).

Off the continental edge, the heat from the surface layer in the summer penetrates more deeply than over the shelf, due to the lower vertical stability of the water column in this area. As a consequence of the higher vertical stability over the shelf area a temperature minimum (< 7 °C) is seen at 50–70 m depth in this area. This is a memento of the previous winter's surface water minimum temperature.

BOX 5.2 MAREANO
– NEW KNOWLEDGE FROM THE NORWEGIAN COASTAL AND OCEANIC AREAS

▶ Deep-water coral reef on 200 m depth in Lopphavet, Northern Norway, 70°30'N. The reefs in Norwegian waters are formed by the stone coral *Lophelia pertusa*. The red gorgonian *Paragorgia arborea* is also a very frequent associate.

Norway has large natural resources in its coastal and shelf regions. Various central, regional and local government bodies manage these. It is a paradox, however, that our information from the surface of the planet Mars is over 100 times as detailed as what is available for the sea bottom off the Norwegian coast, where an important proportion of Norway's economic interests are situated. In 2005, therefore, the MAREANO programme was launched in order to improve this situation. MAREANO is an integrated mapping programme for the Norwegian seas and coastal areas carried out by the Institute of Marine Research (IMR), the Geological Survey of Norway (NGU) and the Norwegian Hydrographic Service (SKSK). The programme includes detailed mapping of the topography as well as the physical, chemical and biological seabed environment on the Norwegian continental shelf and adjacent oceanic areas.

The MAREANO programme will collect and compile knowledge of the coastal region and the shelf in an integrated database and make it available on the Internet using state-of-the-art GIS technology. The goal is to provide society with up-to-date, quality controlled data for management and sustainable development and exploitation.

A clean and healthy environment – processes and environmental baseline
Sediments on the seabed serve as historical archives for the environmental impact of pollutants associated with sediment particles and organic material. MAREANO will provide the necessary knowledge of pollution-related distribution processes, and establish an environmental baseline for the Norwegian Sea.

Safety and framework for economic development – co-operating with industry
The Norwegian Sea is the most promising hydrocarbon province in Norway, but also the most challenging. MAREANO will support the offshore industry by providing a detailed bathymetry and stratigraphy of shelf and slope sediments. Better topographic maps of the sea bottom will mean reduced costs for fisheries and the petroleum industry. Furthermore, disturbance of the seabed may be reduced because improved information will mean that better and lighter fishing gear can be used.

Biodiversity and habitats – linking biology, geology and technology
Knowledge of marine biodiversity and the distribution of marine habitats is a national priority in Norway. Integrating marine biology with marine geology and advanced technology, will improve our knowledge of this area. MAREANO aims to provide a baseline overview of marine habitats in the seas around Norway, and thus a basis for future monitoring of marine biodiversity.

Coral reefs – protected by law, but where are they?
Surprisingly enough, the Mid-Norwegian Shelf has numerous coral reefs. These are cold-water reefs and include the world's largest known reef of its kind – the Sula Reef. The reefs are threatened by several factors, the most serious of which is fishing. MAREANO will provide a detailed database with the locations of the coral reefs, enabling coral reefs to be left undamaged by fisheries and other activities.

The northern Norwegian coast

This part of the coast borders the Barents Sea, which is a shelf sea with an average depth of 230 m. The Norwegian Coastal Current flows close to the coast, and when it crosses the border with Russia it changes its name to the Murman Current. The Norwegian North Atlantic Current along the continental slope of central Norway splits into two off northern Norway, with one branch flowing further north towards Svalbard while the other flows into the Barents Sea as the North Cape Current (Figure 1.1).

The seasonal fluctuations in temperature and salinity of the area are reflected in the observations from the fixed oceanographic station at Ingøy (Figures 5.8 and 5.12). The average long-term minimum temperature in the deeper layer is about 4.5 °C and occurs nearly simultaneously at all depths down to 300 m in March–April. The maximum temperature of just above 9 °C is found in the surface layer in August. This seasonal extreme also propagates downwards, taking about two months to reach 300 m. The propagation time for the seasonal extreme to reach the deeper layer is approximately half the time needed off the coast of mid-Norway. This is a result of the northerly decrease in the vertical stability of the water masses.

The long-term minimum surface layer salinity of just above 34 is observed in August and this reaches a depth of 300 m in December. The maximum surface layer salinity of around 34.4 is recorded in May, while at depths below 200 m it is observed in September.

Figures 5.10 and 5.11 show the long-term mean vertical distribution of temperature and salinity in the Fugløya–Bjørnøya section (Figure 5.8). During the summer, Atlantic Water, with a salinity of more than 35, is raised higher in the water column, and the low-salinity waters of the surface layer are displaced further offshore.

Temporal and spatial distribution of nutrients

Francisco Rey, Jan Aure and Didrik S. Danielssen

6.1 The role of nutrients in Norwegian oceanography

The term "nutrients" refers specifically to those chemical components present in seawater that are of direct significance for plant growth. In coastal waters nutrients are consumed by benthic algae and by phytoplankton. Therefore, in monitoring and scientific programmes, nutrients such as nitrate, phosphate and silicate are usually measured in order to evaluate the natural productivity of an area or to estimate its degree of eutrophication in cases of non-natural increases, such as transport from land through freshwater runoff.

With respect to the Norwegian Coastal Current there are several sources of nutrients including the fjords that transport urban wastewater and industrial discharges; some also comes from agriculture, although this is much lower in quantity than what is transported from the continent. Input from adjacent sea areas is also important for advective transport of nutrients to the coastal current. The main contributions at the origins of the Norwegian Coastal Current come from the southern North Sea through the Jutland Current that flows along the west coast of Denmark and into the Skagerrak, the Baltic waters with further additions from the Kattegat and Belt Sea and the coastal areas of West Sweden and Southern Norway. In the Skagerrak area and further north along the coast of Western Norway, the admixture of Atlantic Water is also an important source of nutrients.

Nutrient measurements, with some restrictions, can also be employed to identify and partially quantify certain aspects of different water masses. In coastal areas close to inhabited or industrial regions, there is often a close relationship between types and concentrations of nutrients and the salinity of the water masses, in some cases enabling these to be more accurately identified.

Although the importance of nutrients for phytoplankton growth in the sea had already been etablished by the end of the 19th century, its routine measurement did not start until after the First World War. These measurements were mainly intended to provide ancillary information to scientific studies that centred on phytoplankton productivity, with the aim of evaluating potential biological productivity in coastal and oceanic waters. The incorporation of nutrients in hydrographic programmes as separate parameters did not start until it was discovered that, in many coastal areas, discharges of nutrients from urban and agricultural regions led to the eutrophication of these areas. The first Norwegian area to be recognized as presenting this kind of problem was the inner Oslofjord. Water quality studies carried out by University of Oslo scientists indicated that sewage discharges raised phytoplankton production levels through the release of nitrate and phosphate from the decomposition of organic matter. At the time this phenomenon was considered to have a positive effect on the productivity

BOX 6.1 THE CHEMICAL COMPOSITION OF SEAWATER

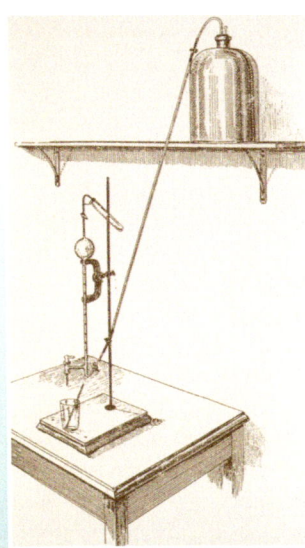

▶ The Knudsen burette, which was widely used since its introduction in the early 1900s and until the 1960s to determine the salinity of seawater by titration with silver nitrate. Here, the set-up used on the Norwegian Research Vessel "Michael Sars" on its expedition to the North Atlantic in 1910.

▶ The Norwegian Research Vessel "Michael Sars" which carried out extensive studies of northern European waters at the beginning of the 20th century.

Seawater is a complex solution with a great diversity in its chemical composition. Probably all known chemical substances are present in seawater, either as elemental molecules and/or suspended matter, but most of them occur in extremely small proportions. The total mass of all solid substances dissolved in one kilogram of seawater is called the salinity and it is expressed as parts per thousand. The modern definition of salinity has changed according to the methodology used in its measurement. Today, salinity is not longer measured by chemical analysis but by means of conductivity sensors and its expression is dimensionless.

The two major elements in seawater are obviously hydrogen and oxygen. These two elements alone make up 96.5 % of the water in weight while the rest of all the dissolved elements make up only 3.5 %. As pointed out above this is the salinity of the seawater. Eight main chemical elements account for about 99.66 % of the total weight of these dissolved substances (see Table).

Nutrients, as different forms of nitrogen, phosphorus and silicon, make up only about 0.05 % of the total dissolved substances (see Table). However, due to their utilization by microorganisms, they are present in widely varying concentrations in time and space and are of vital importance for the biological productivity of the oceans.

The British chemist Robert Boyle (1627–1691) was the first to publish a chemical treatise on the sea in 1674. However, like other chemists such as Lavoisier, Bergman, Murray and others, he was primarily interested in the analysis of seawater because of the presence in it of so many chemical components. However, the first to show a specific interest in seawater specifically in terms of its chemical composition with regard to its physical properties, was the French chemist and physicist Joseph Louis Gay-Lussac.

The average salinity of the oceans of 35 was first roughly obtained by Gay-Lussac who, analysing samples from the Atlantic Ocean, found an average "saline residue" of 3.65 g/kg (today's salinity of 36.5). However it was the German chemist William Dittmar who in 1884, on analysing 77 sea water samples from many parts of the world collected during the scientific expedition of the British corvette HMS Challenger (1872 to 1876) found that the average world ocean salinity was about 35. One of the main conclusions drawn from Dittmar's work was that the proportions of the major constituents of seawater are almost constant throughout the world. His average concentrations are still used today to represent the ratios of the major constituents.

▶ Average concentrations (g/kg) and percentages of the total elemental content of some of the major dissolved chemical components in seawater (S=35) as well as the main nutrient elements.

Component	g/kg	Main form	%
Chlorine (Cl)	19.3529	Cl^-	54.88
Sodium (Na)	10.7596	Na^+	30.52
Sulphur (S)	2.7124	SO_4^{2-}	7.69
Magnesium (Mg)	1.2965	Mg^{2+}	3.68
Calcium (Ca)	0.4119	Ca^{2+}	1.17
Potassium (K)	0.3991	K^+	1.13
Carbon (C)	0.1412	HCO_3^-	0.40
Bromine (Br)	0.0673	Br^-	0.19
Sum	**35.1409**		**99.66**
Nitrogen (N)	0.01550	NO_3^-	0.04
Phosphorus (P)	0.00009	PO_4^{3-}	<0.01
Silicon (Si)	0.00290	SiO_4^{4-}	0.01

of the fjord. It was after the Second World War that other scientists from the same university found that the fjord was suffering from an enormous increase in phytoplankton production that adversely affected marine life. The decomposition of the large amounts of organic matter formed by photosynthesis led to oxygen deficiency in many areas. The result was that marine life on the seabed was greatly reduced.

At the beginning of the 1960s, a nutrient monitoring programme in the Oslofjord was established; the first of its kind in Norway. The main objective of the programme was to determine what was causing the wastewater pollution and to understand the processes involved and their effects on marine life in the fjord. The results rapidly confirmed that the water quality of the fjord was powerfully affected by the nutrients in the wastewater. These results also acted as the trigger for the start, in 1975, of a nutrient and oxygen monitoring programme in 27 Norwegian fjords. This programme, carried out by the Institute of Marine Research, is still running today.

The first synoptic measurements of nutrients on the Norwegian coast were carried out in the mid-seventies (Føyn and Rey, 1981) in connection with various scientific programmes, but routine analysis at the oceanographic sections did not start until the 80s in the North Sea area and the 90s from the west coast of Norway and northwards. Nowadays, nutrients measurements are incorporated into the IMR's oceanographic monitoring programme, which covers several sections across the Norwegian Coastal Current several times a year.

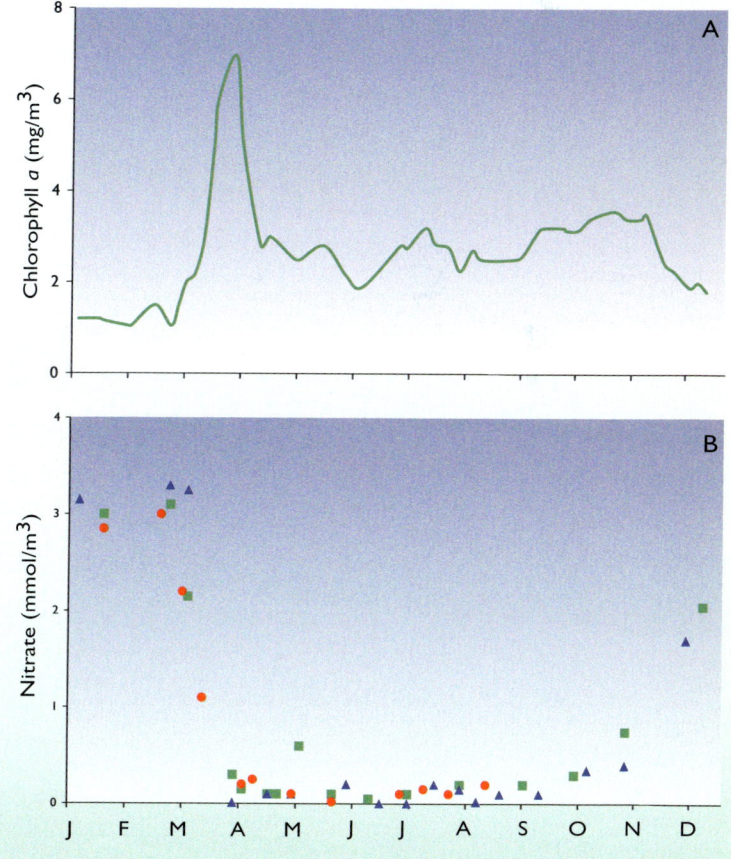

▶ Figure 6.1
Annual variability in the distribution of chlorophyll *a* and nitrate in surface waters of the Arkona Basin (Baltic Sea).
A) Weekly averages of chlorophyll *a* in mg/m^3 for 1992–2005.
B) Nitrate concentrations in mmol/m^3 for 2004 (green squares), 2005 (blue triangles) and 2006 (red dots).
(Modified from Maunula 2006a and 2006b).

6.2 Sources of nutrients in the Norwegian Coastal Current

As pointed out in Chapter 4, the Norwegian Coastal Current has its most important origins in the outflow from the Baltic Sea (about 50 %), with further contributions coming from the freshwater runoff to the Norwegian coast (40 %) and from waters from the North Sea (10 %). As it spreads northwards, oceanic Atlantic Water also affects the Coastal Current by lateral mixing as the two bodies of water run parallel along the Norwegian coast, resulting in a gradual northward increase in salinity. Its chemical properties thus tend to vary according to the degree of mixing and the proportions supplied by the different water masses. Furthermore, the distribution of nutrients will be affected by the biological activity and will display strong seasonal and regional variability.

Although the outflow from the Baltic Sea makes a significant contribution to the formation of the Norwegian Coastal Current, this is not reflected to the same

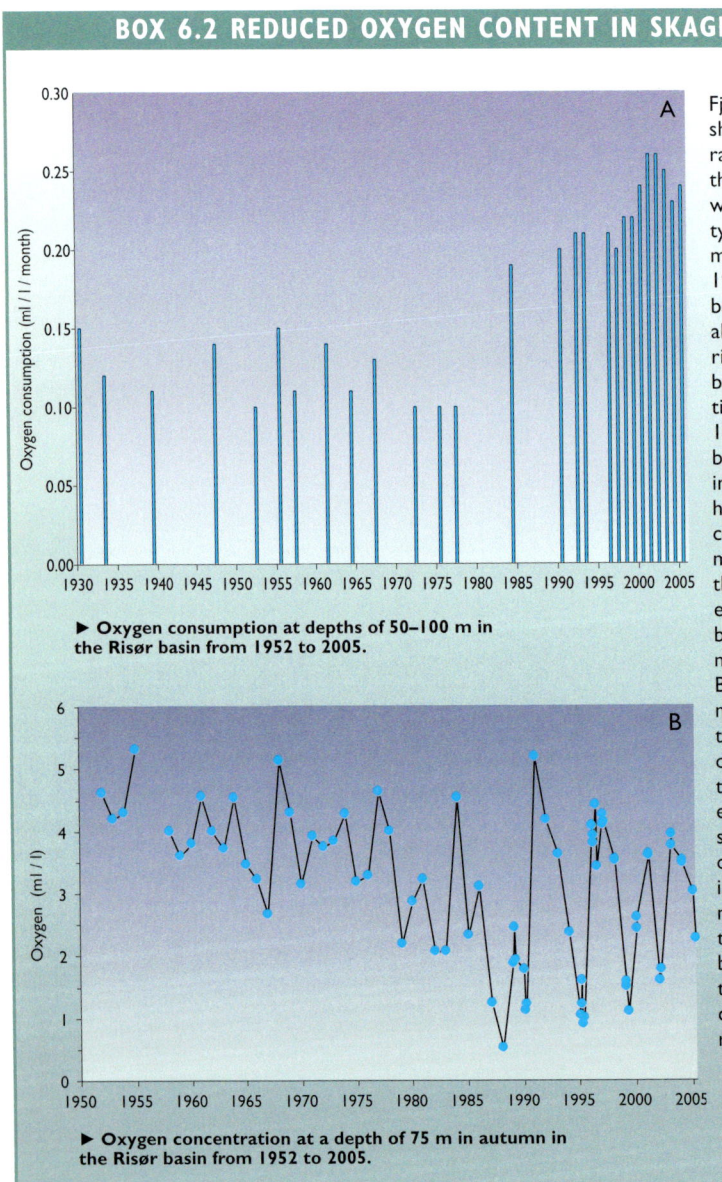

BOX 6.2 REDUCED OXYGEN CONTENT IN SKAGERRAK FJORDS

▶ Oxygen consumption at depths of 50–100 m in the Risør basin from 1952 to 2005.

▶ Oxygen concentration at a depth of 75 m in autumn in the Risør basin from 1952 to 2005.

Fjord basins are characterised by shallow sills, which reduce the rate of water exchange between the basin water and the coastal water outside the fjord. There is typically a large supply of organic material to the fjord. Since about 1980 oxygen consumption in the basins of the Norwegian fjords along the Skagerrak coast has risen considerably. In the Risør basin, for example, consumption has almost doubled since 1980 compared to the period before 1980 (Figure A). This increase in oxygen consumption has resulted in lower minimum concentrations of oxygen in many of the fjordic sill basins on the Skagerrak coast. The greatest reductions occurred in sill basins with natural low oxygen minimum concentrations (Figure B). Long-term observations of nutrient and oxygen concentrations suggest that the rise in oxygen consumption is related to the increased large-scale eutrophication in the Skagerrak since about 1980. The poorer oxygen conditions have resulted in a drastic reduction in the numbers of species living close to the bottom in many fjord basins on the Skagerrak coast. In the fjords with the worst oxygen conditions, biodiversity has been reduced by as much as 50–90 %.

extent in the levels of nutrient concentrations found in the Coastal Current. This is due to the relatively lower nutrient concentrations in the waters flowing out from the Baltic Sea through the Arkona Basin and into the Kattegat compared with the other sources of water. Figure 6.1 shows the annual variability in chlorophyll *a*, an index of phytoplankton concentration, and nitrate concentrations in the Arkona Basin. The relatively high concentrations of phytoplankton at this location throughout the year keeps the nitrate concentrations low compared to, for instance, the southern North Sea, where nitrate concentrations can reach values up to ten times as high. Most of the nutrients in this latter region have their origin in the freshwater runoff from the large European rivers. These water masses are mostly transported into the Skagerrak by the Jutland Current and later become mixed with central North Sea Water to contribute to the formation of the Norwegian Coastal Current. Only occasionally do these waters enter the Kattegat region (Ærtebjerg et al., 2003).

Figure 6.2 shows the horizontal distribution of salinity and nitrate at a depth of 10 m in the North Sea during late autumn 1994. The salinity of the Baltic outflow at this time of the year is below 24, while nitrate concentrations are about 1 mmol/m^3. On the other hand, the inflow of North Sea waters into the Skagerrak along the Danish coast displays salinities of between 32 and 34 and nitrate concentrations between 4 to 6 mmol/m^3. It is also important to emphasize that the inflow of North Sea waters is not a continuous process but rather take place as pulses, occasionally bringing water masses with larger nitrate concentrations than those observed in this particular example. Atlantic Water (S> 35) on the western side of the Norwegian Coastal Current also displays larger concentrations of nitrate than those from the Baltic outflow. Through mixing, this water mass makes an important contribution to the increase in nutrient concentrations in the Coastal Current, especially from the southern part of Norway and northwards. The Atlantic Water also enters the deep trench in the Skagerrak area and, by mixing from below, helps to increase nutrient levels in the Norwegian Coastal Current at its origins.

As nutrients are utilized in the building of biological material, their seasonal concentration is powerfully regulated by phytoplankton growth. At the same time, phytoplankton growth is regulated by light conditions and the physical oceanography of a particular region. Physical stratification of the upper layers, either by low salinities or higher temperatures, and shallow areas are the most

▶ **Figure 6.2**
Horizontal distribution of salinity (left panel) and nitrate (mmol/m^3) (right panel) at a depth of 10 m in the North Sea and adjacent areas in late autumn 1994.

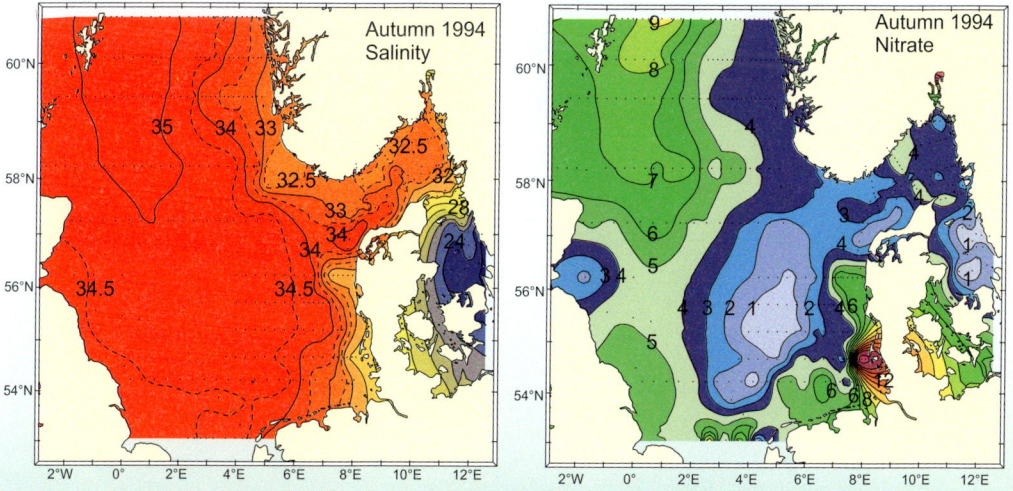

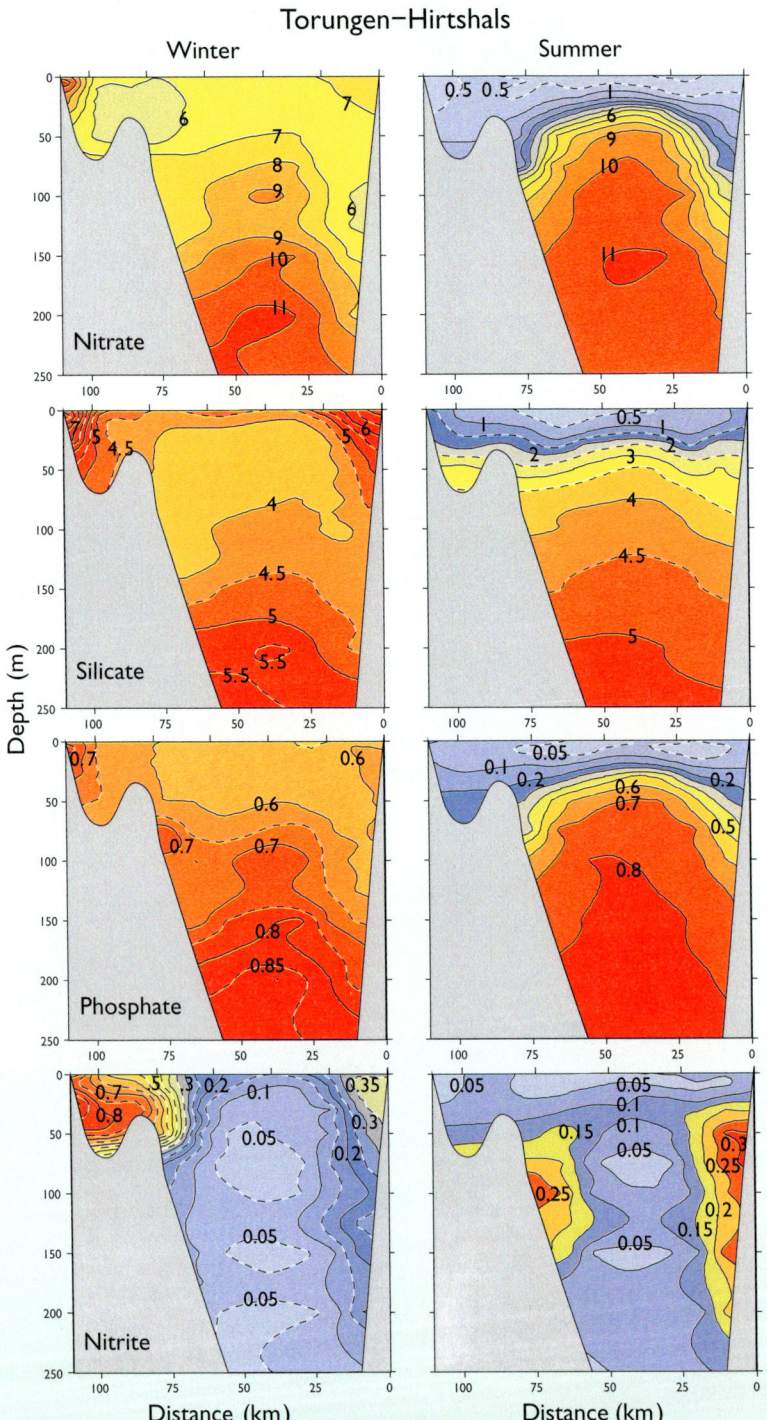

▶ Figure 6.3
Vertical distribution of average nitrate, silicate, phosphate and nitrite concentrations (all in mmol/m³) during winter (January) and summer (August) along the Torungen–Hirtshals section in 1982–2005.

important factors in keeping phytoplankton in the upper water column, where they can take advantage of the better light conditions to grow. In the case of the Arkona Basin, for example, phytoplankton concentrations are at relatively high values from March to December, keeping nitrate concentrations low.

6.3 Mean geographical variations in nutrients

The hydrographical conditions along the Norwegian Coastal Current change from south to north, as discussed in Chapter 5. This will undoubtedly also be reflected in the distribution of nutrients. Three oceanographic sections representing three main areas along the Norwegian coast have been selected to illustrate this situation: the Torungen–Hirtshals section for the Skagerrak area, the Svinøy section for the central Norwegian coast and the Fugløya–Bjørnøya section for the northern Norwegian coast (for location see Figure 5.8). In order to increase the vertical resolution in the figures, the depth of the sections has been limited to 250 metres, as this depth covers most of the water masses belonging to the Coastal Current.

Figure 6.3 shows the mean distribution of nutrients in winter and summer across the Skagerrak from the Danish coast on the left, to the Norwegian coast. The distribution is based on average values for 1982–2005. During the winter, as would be expected, nutrient concentrations in the upper layers are higher than during the summer, especially on the coastal sides. In the central part of the section, nutrient levels are closer to those found in Atlantic Waters, reflecting the lower stratification that leads to the winter vertical mixing and replenishment of nutrients from deeper layers. The dome shape of the nutrilines reflects the hydrography of this area, as discussed in Chapter 5. On the Danish side, even higher levels can be observed, indicating the presence of waters from the eastern North Sea that flow into the Skagerrak via the Jutland Current. This pattern is especially marked in the nitrite distribution. On the Norwegian side of the section, nutrient levels are also high, although they are somewhat lower than on the Danish side.

During the summer, nutrients in the upper layer reach their lowest values. In the central part of the section the sharp gradients observed between the upper and deeper layers makes the dome-shaped nutrilines much more marked than in winter. Nutrient depletion also reaches greater depths on both coastal sides than in the central part, further emphasising the domed shape. In contrast to the winter situation, nitrite during summer shows a completely different distribution on both coastal sides. Areas of high nitrite are clearly shown below 30–50 m, and extend to depths of 150–200 m. While the high-nitrite area on the Danish side observed during winter seems to be a result of transport by surface currents, the subsurface areas during summer are most probably due to the accumulation of nitrite due to biological processes that take place close to the bottom. Both these summer areas are situated in a region mostly dominated by the transition between the Coastal and Atlantic Waters (see Figure 5.11). The core of Atlantic Water in the central part of the section shows very low nitrite levels in both winter and summer.

Figure 6.4 shows typical examples of the vertical nutrient distribution at the Svinøy section. In this section the Norwegian Coastal Current runs northwards parallel to the Norwegian Atlantic Current. The separation between these two water masses can easily be observed in the salinity and temperature distribution (see Figures 5.10 and 5.11). During both the winter and summer, the Coastal Water spreads above the Atlantic Water in a wedge shape that is narrower and deeper during the winter. The same general pattern is observed in the nutrient distribution, especially during the winter situation (January 1998). A sharp, almost vertical gradient in nitrate, silicate and phosphate separates the two water masses. The Coastal Water is also characterized during the winter season by nutrient levels that are about 30 % lower than that of the Atlantic Waters (i.e. 6–8 mmol/m^3 nitrate as against >11 mmol/m^3). The nitrite distribution also follows the same pattern, although this is not as distinct as it is in the three other classes of nutrients.

During the summer (July 1999), the nutrient distribution is somewhat different than in winter. In addition to the spread of Coastal Waters off the coast and above the Atlantic Water, the consumption of nutrients by phytoplankton creates a sharp horizontal gradient at a depth of between 30 and 50 metres across the section, with lower values in the upper layer. In this season the nitrite distribution also shows a different pattern than the other nutrients, with higher values at the nutricline (the nutrient gradient) in the region covered by Atlantic Water. These higher levels are the product of nitrite produced by biological activity. The reason that this accumulation is not observed above the continental shelf, in the Coastal Water, is that the biological season here starts earlier than in Atlantic Water and most of the nitrite have disappeared, mainly by mixing. The main reason for this difference in time for the biological activity is the presence of a deeper upper mixed layer in the Atlantic Water due to its more homogeneous vertical distribution in temperature and salinity, and therefore density. This results in earlier phytoplankton growth in the Coastal Water since phytoplankton remains closer to the surface and under better light conditions than in the Atlantic Water. The biological activity in the upper layers of the Atlantic Water starts when the warming of this layer creates a stronger stratification.

Another characteristic feature during summer is the more landward distribution of water masses, with high nutrients levels below the Coastal Water. This is due to the intrusion of Atlantic Water over the continental shelf in compensation for the offshore spread of Coastal Water in the upper layers.

In northern Norway, the vertical distribution of nutrients (Figure 6.5) is represented by observations made at the Fugløya–Bjørnøya section. This section is regarded as representing the transition between the Norwegian Sea and the Barents Sea, although the boundary is largely a matter of convention. The distribution patterns are quite similar to those found at the Svinøy section. One important difference that needs to be borne in mind between the two sections is the topography of the regions. While the Svinøy section extends from the coast through the continental shelf and further west over the deep abyss of the Norwegian Sea, the Fugløya–Bjørnøya section is situated for the most part over the continental shelf at the western edge of the Barents Sea with depths shallower than about 490 m. This is reflected in the weaker gradient between Coastal and Atlantic Water at the Fugløya–Bjørnøya section, especially during the winter season, as well as in the more seawards position of the Atlantic Water. This pattern is the result of a greater degree of mixing of the two water masses in this region, a fact that can also easily be observed in the nutrient distribution, where there is a gradual transition from the Coastal Water, with its lower nutrient levels (for instance 7–8 mmol/m^3 nitrate), to the Atlantic Water with higher levels (> 11 mmol/m^3 nitrate). The exception, as in the Svinøy section, is the nitrite, which are low along the whole section and at all depths.

During summer a strong nutricline is observed along the section at depths of between 30 and 100 m. This nutricline is deepest and weakest close to the coast, and it becomes shallower and stronger offshore following the main pat-

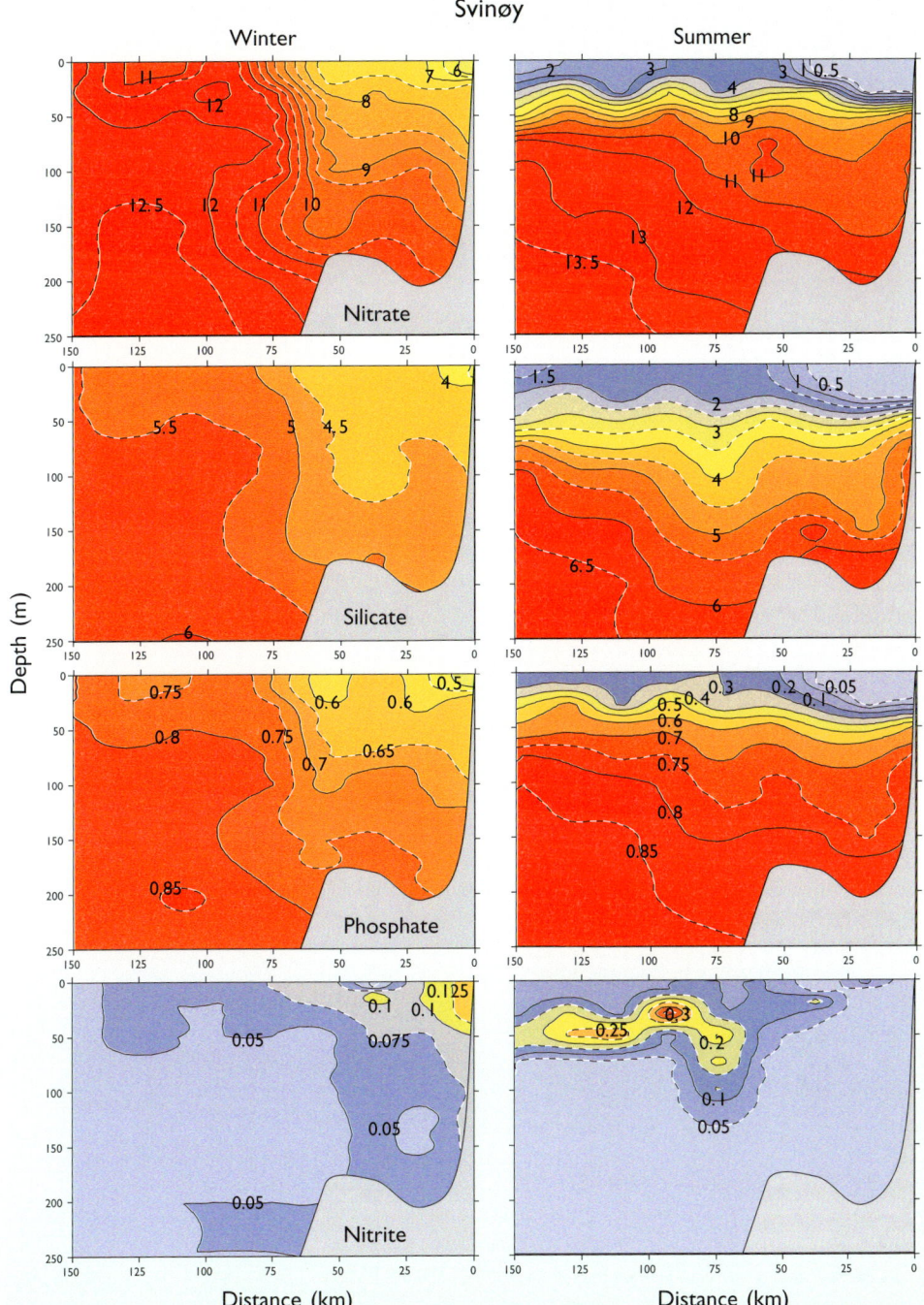

▶ **Figure 6.4**
Vertical distribution of nitrate, silicate, phosphate and nitrite (all in mmol/m³) during the winter (January 1998) and summer (July 1999) along the Svinøy–NW section.

Temporal and spatial distribution of nutrients 81

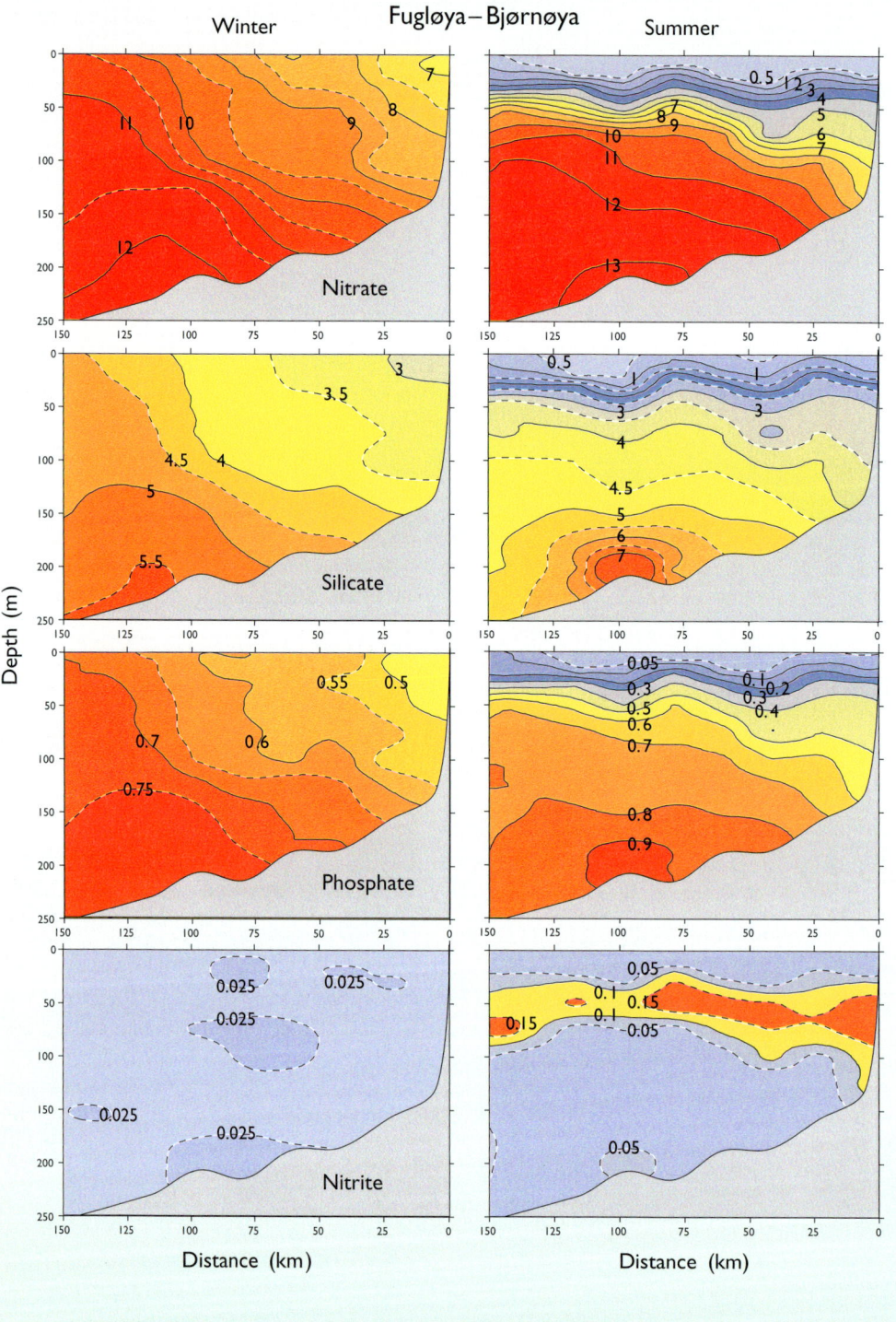

▶ Figure 6.5
Vertical distribution of nitrate, silicate, phosphate and nitrite (all in mmol/m^3) during winter (March 2001) and summer (July 1999) along the Fugløya–Bjørnøya section.

tern of the hydrography of the area. Nitrite concentrations during the summer are low along the section, with the typical accumulation at the nutricline due to biological activity.

6.5 The mean seasonal variation of nutrients

Nutrient levels in the upper layers are greatly affected by their utilization by phytoplankton, free-floating microorganisms, that via photosynthesis, utilize them to build the organic matter that forms the basis of all oceanic life. Undoubtedly then their distribution will show a strong seasonal variability. Furthermore, the nutrient pattern distribution will also reflect to a certain extent the type of phytoplankton found in a given water mass. Three main groups of phytoplankton are considered to be the most important in the ocean: the diatoms, the flagellates and the dinoflagellates. All of them utilize nitrate and phosphate, but the diatoms also need silicate to build up their silicaceous frustule (external skeleton). Diatoms are generally considered to be the most important food for zooplankton in most of the ocean, and as such, they are the basis of many food webs of economic importance.

The seasonal distribution of nitrate at four stations at the three selected oceanographic sections is shown in Figure 6.6. The upper panel in Figure 6.6 shows the average monthly nitrate distribution at one station on the Torungen–Hirtshals section close to the Danish coast in 1982–2005. This region is relatively shallow, and its hydrography is powerfully affected by physical forcing such as currents and tides, resulting in little stratification in the upper 50 metres. This is also reflected in the distribution of nitrate, which, with exception of a winter period between February and March, when it is possible to observe a pulse of water with the highest nitrate concentrations near the surface, is quite homogeneous throughout the water column. The nitrate concentration diminishes rapidly from April to values that are only half of those observed in winter in the course of about 15 days. After that there is a more gradual disappearance throughout the summer until September when it begins to increase again homogeneously throughout the water column. This autumnal increase is primarily a result of the nutrient replenishment that occurs when phytoplankton, due to less optimal growth conditions, are unable to consume nitrate as fast as it is regenerated. Replenishment continues through the winter and is homogeneous through the whole water column.

The seasonal distribution of nitrate at the northern end of the section, close to the Norwegian coast, is shown in the second panel of Figure 6.6. The general pattern of distribution is similar to that on the Danish side, but in winter, nitrate do not reach the same high values and it is fairly homogeneously distributed down to the bottom. The disappearance of nitrate in the upper layers starts about one month earlier than on the Danish side and reaches very low values much more rapidly. From April to October there is almost no nitrate left in the upper 10–20 metres. Moreover, because the station is deeper, nitrate is gradually consumed to depths to about 100 m during the summer before they once again begin to be replenished in the autumn by mixing with water from deeper layers rich in nitrate.

The seasonal distribution of nitrate off the west coast of Norway is represented by the observations at the Svinøy section (third panel in Figure 6.6). Here too, the general pattern is similar to that in the northern Skagerrak region, but with two main differences. First, the winter vertical mixing reaches only as far down as depths of 50–75 m, and nitrate concentrations diminish rapidly with

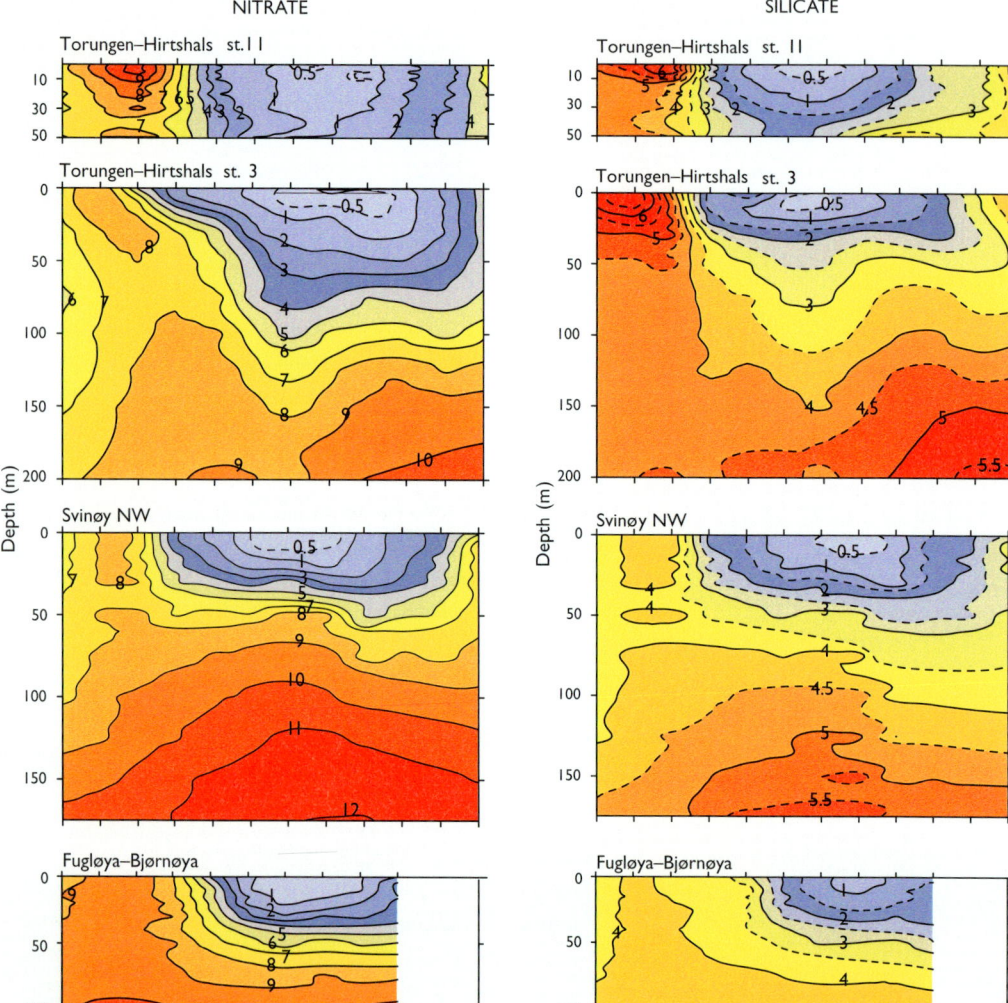

▶ **Figure 6.6**
Seasonal distribution of average nitrate (mmol/m³) at selected stations in the Norwegian Coastal Current:
St. 11 Torungen–Hirtshals section on the Danish side (period 1982–2005);
St. 3 Torungen–Hirtshals section on the Norwegian side (period 1982–2005);
St. 2–4 Svinøy–NW section (period 1990–2004) and St. 2–4 Fugløya–Bjørnøya section (period 1980-2005).

▶ **Figure 6.7**
Seasonal distribution of average silicate (mmol/m³) at selected stations in the Norwegian Coastal Current:
St. 11 Torungen–Hirtshals section on the Danish side (1982–2005);
St. 3 Torungen–Hirtshals section on the Norwegian side (1982–2005);
St. 2-4 Svinøy–NW section (1990-2004) and St. 2–4 Fugløya–Bjørnøya section (1980–2005).

time from March onwards, down to about 30–40 m, subsequently stabilizing at this depth without major further consumption deeper in the water column, as in the northern Skagerrak area. The greatest influence on the vertical distribution of nitrate at this section by the Atlantic Waters is a result of the higher concentrations below 50–75 m throughout the year. A similar pattern of distribution can be observed in the northernmost section of Fugløya–Bjørnøya, with the main difference that the period of nitrate disappearance commences later than at the southern sections and is of a shorter duration (lower panel Figure 6.6). There are no routine observations at this section in November and December.

The seasonal distribution of silicate at the three sections is despicted in Figure 6.7. The main patterns of distribution are quite similar to those from nitrate for all examples. One main difference is during spring (March–April) at the Fugløya–Bjørnøya section (Figure 6.7 lower panel) where silicate diminishes at a much lower rate with time than nitrate. This fact can be interpreted by the presence of a higher proportion of flagellates with respect to diatoms during this period and thus a relative higher consumption of nitrate with respect to silicate.

6.6 Anthropogenic influence on the nutrient concentrations in the Skagerrak Coastal Water

The Norwegian Coastal Current along the Skagerrak coast mainly transports the Norwegian Skagerrak Coastal Water (NSCW) in the upper 30 m. This water mass is a mixture of water flowing into the area from the North Sea by the Jutland Current and the Baltic outflow from the Kattegat transporting the Kattegat Surface Water (KSW) (Figure 8.1). The Jutland Current consists of the German Bight Water (GBW) and Southern/Central North Sea Water (S/CNSW) (Aure et al., 1998). In the winter and spring the average contributions of KSW, S/CNSW and GBW to the Skagerrak Coastal Water in the upper 30 m are 25, 55 and 20 %, respectively (Figure 6.8).

The supply of anthropogenic nitrogen to the Katttegat and the German Bight increased significantly in the late 1970s (Aure et al., 1998). In the German Bight the average nitrate concentration during the period January to May approximately tripled, in the course of a few years, from 14 to 40 mmol/m^3. Phosphate, however, showed a slight decrease so the N:P ratio during the same period rose from 19 to 54. In the Katttegat, the average winter concentration of nitrate rose from 5 to 7 mmol/m^3 and phosphate from 0.6 to 0.7 mmol/m^3, while the Southern/Central North Sea Water showed insignificant changes after about 1980.

Reflecting the differences in nutrient concentrations, the three main sources of water contribute to different fractions of the nutrient content in the Skagerrak Coastal Water (Figure 6.8). The mean winter–spring contribution of nitrate to the Skagerrak Coastal Water from the German Bight is estimated at 75 %, while the contribution from the Kattegat Surface Water is estimated at 11 %. This is evidence that in an average year the inflow of German Bight Water via the Jutland Current is the dominant source of nitrate to the Skagerrak Coastal Water (0–30 m). The average contributions of phosphate in winter–spring from the German Bight and the Katttegat are 39 % and 28 %, respectively.

In the upper 30 m of the Norwegian coastal water off Torungen, the mean winter–spring concentrations

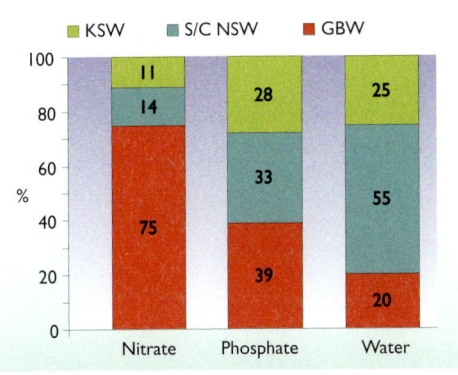

▶ Figure 6.8
Mean percentage contributions of Kattegat Surface Water (KSW), Southern/Central North Sea Water (S/CNSW) and German Bight Water (GBW) to the Skagerrak Coastal Water (0–30 m) during the winter and spring, in terms of fraction of water mass, nitrate and phosphate concentrations since about 1980.

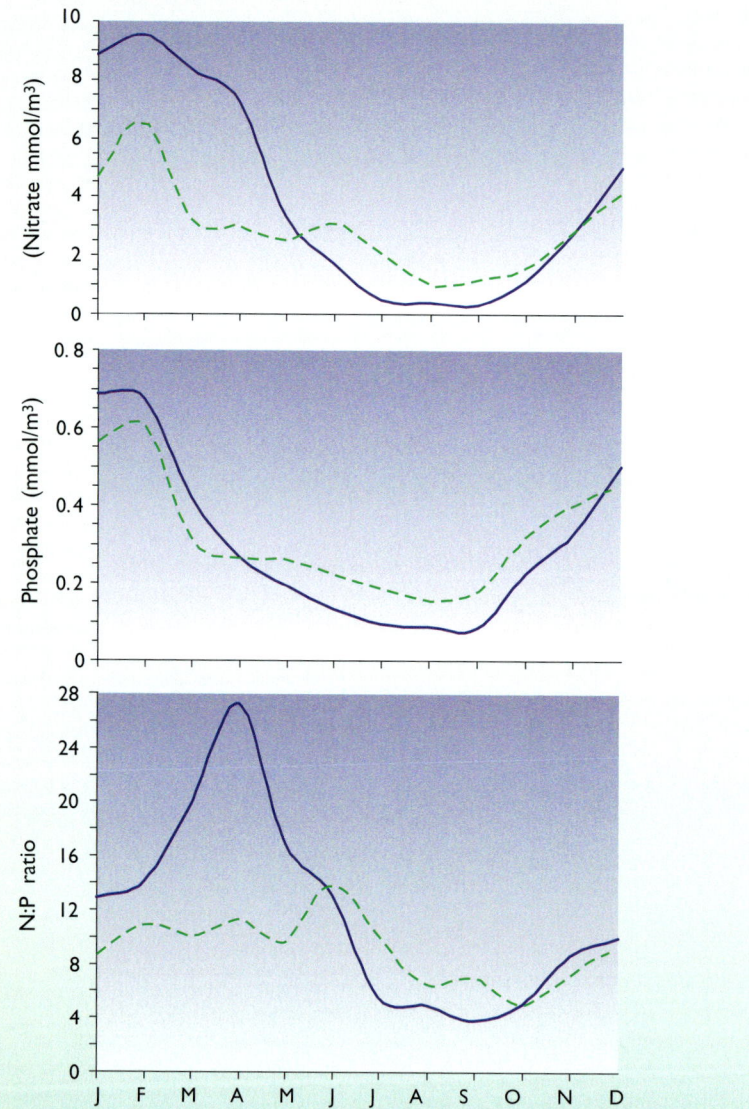

▶ **Figure 6.9.**
Monthly means of phosphate, nitrate and the N:P ratio in Skagerrak Coastal Water (0–30 m) off Arendal during 1975–1980 (stippled green line) and 1990–1995 (solid blue line).

of nitrate were considerably higher in the 1990s than between 1975 and 1980 (Figure 6.9). As explained above, this increase largely reflects the increased anthropogenic supply of nitrogen to the German Bight and Kattegat since the late 1970s. A maximum increase of 160 % (3 to 8 mmol/m^3) appeared in March, while the average increase over the winter–spring period has been about 100 %. As expected, changes in phosphate have been insignificant since 1980 (Figure 6.9). The imbalance in the supply of anthropogenic nutrients to the Norwegian Skagerrak Coastal Water led to increased N:P ratios in the winter–spring period, with the greatest increase in April from 11 to 28 (Figure 6.9). The low nitrate and phosphate concentrations between May and October are a result of increased consumption by benthic algae and phytoplankton. During this growth period nutrients are mainly incorporated into algae, phytoplankton and organic particles.

Interannual variability in the freshwater runoff has a significant impact on nitrate concentrations in the upper layers both in the Katttegat and in the German

BOX 6.3 ARTIFICIAL UPWELLING IN FJORDS

Nutrient availability is a necessary condition for growth of "the grass of the sea", the phytoplankton which in turn determine the rate of production at higher trophic levels. Both natural and anthropogenic nutrient enrichment are capable of having positive effects in terms of increased biological production. The great fisheries along the coasts of Chile and Peru and off northwest Africa are due to natural wind-induced upwelling of nutrient-rich water from deeper ocean layers to the photic zone. This transport of nutrients to the upper layer results in a rich plankton production and supply of food for fish.

In 2004 and 2005 a large-scale artificial upwelling experiment was carried out in a Norwegian fjord. Brackish water from the surface layer was pumped down to 30 m (Figure A). Due to its lower density than the surrounding water, the brackish water ascends to the photic zone. During its rise, the brackish water carries with it about 10–15 times its own volume of nutrient-rich deep water to the upper layer, thus providing nutrients for phytoplankton growth. In the course of the artificial upwelling, which took place in the summer, the algal biomass, expressed as chlorophyll a, approximately tripled inside an area of influence of about 10 km² (Figure B). Artificial upwelling has the potential to form the basis of an enhanced and more predictable mussel cultivation industry in oligotrophic Norwegian fjords.

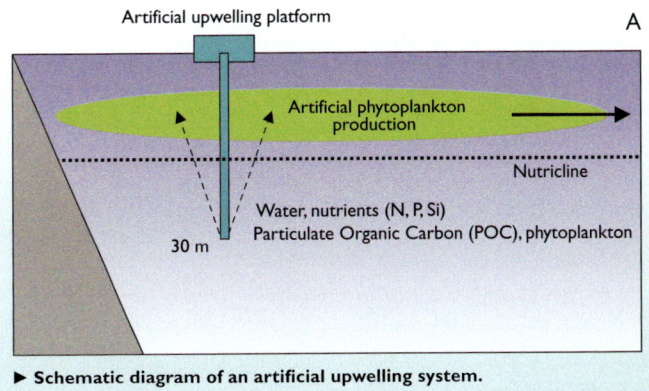

▶ Schematic diagram of an artificial upwelling system.

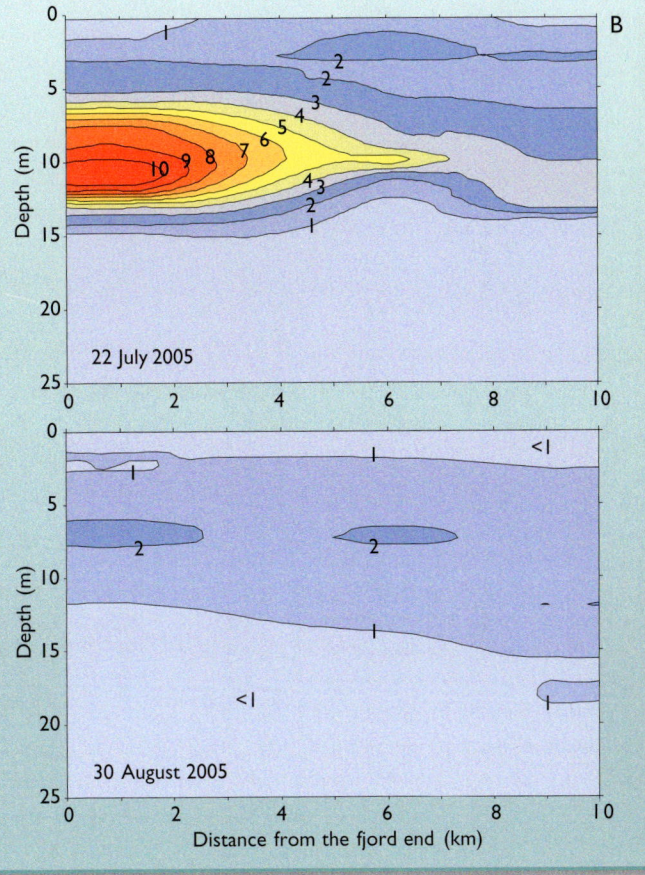

▶ Vertical distribution of chlorophyll a (mg/m³), from 0 to 25 m in the inner 10 km of the Lysefjord. Upper panel: artificial upwelling on 22 July 2005. Lower panel: natural conditions on 30 August 2005.

Bight, while the impact on phosphate and silicate concentrations is much less. The excess of nitrogen in current anthropogenic inputs is then reflected in higher N:P and N:Si ratios, particularly in years with high freshwater runoff. In 1994 there was very high freshwater runoff from continental rivers and at the end of April this year the highest nitrate concentrations ever observed in the German Bight appeared (about 100 mmol/m³). In contrast, after a relatively dry winter, such as in 1996, the nitrate concentration fell to about 30 % of the values in years

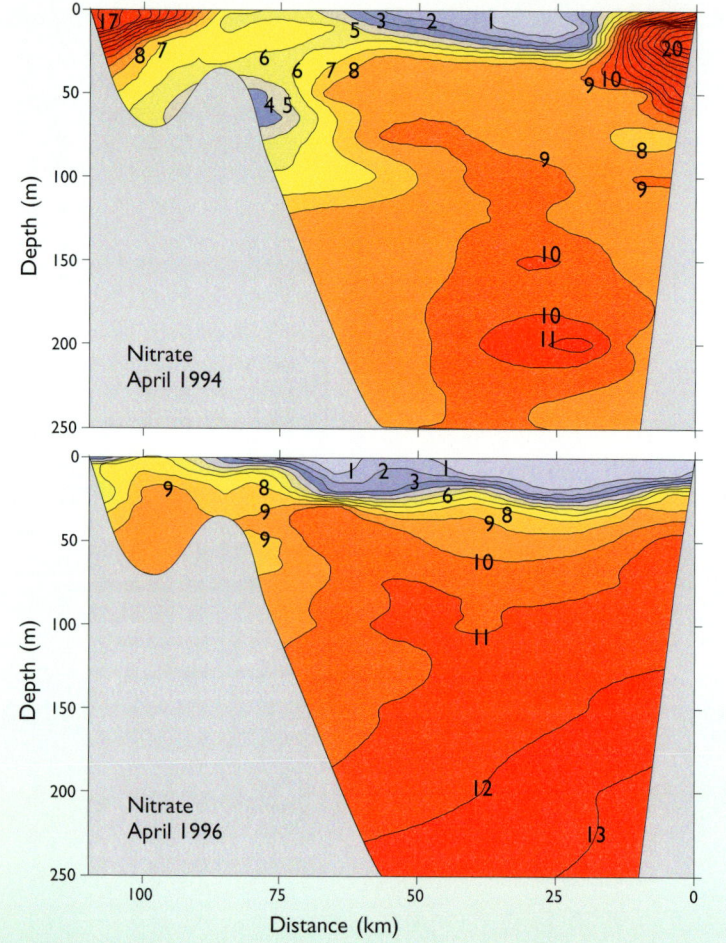

► **Figure 6.10**
Nitrate concentration (mmol/m³) in the Torungen–Hirtshals section for April 1994 and April 1996.

with high freshwater runoff. The high runoff of anthropogenic nitrate to the German Bight in 1994 had a significant impact on nitrate concentrations in the Skagerrak Coastal Water. Figure 6.10 shows a maximum nitrate concentration of about 30 mmol/m³ in the inflowing Jutland Coastal Water off Hirtshals and about 22 mmol/m³ in the Skagerrak Coastal Water off Torungen (Arendal) in April 1994. In contrast, the maximum concentrations in the Skagerrak Coastal Water at a depth of 20 m off Torungen in April 1996, after a relatively dry winter, were 8–9 mmol/m³ (Figure 6.10).

The elevated nutrient supply and concentrations since about 1980 have increased primary production in the Skagerrak Coastal Water and the supply of plankton and organic matter from the Kattegat and the southeastern North Sea. The nutrient enrichment and the increased N:P ratio may also have brought about an increase in the occurrence and risk of harmful algae species. One consequence of the nutrient enrichment is the elevated oxygen consumption in fjord basins along the Skagerrak coast since about 1980 (see Box 6.2).

Short-term variability and small-scale features

Roald Sætre

7.1 Introduction

The temporal variability in the properties and dynamics of the ocean spans a wide range of scales – from seconds to centuries. Chapter 5 deals with the long-term mean situation of coastal water-mass properties, their spatial variability and the mean seasonal fluctuations. Inter-annual variations up to decadal scale or so-called ocean climate fluctuations are considered in Chapter 10. In this chapter we will deal with short-term temporal variations at scales ranging from days to weeks, where meteorological forces such as wind are the main driving force.

A current map with many arrows as seen in Chapter 2 (e.g. Figure 2.12) depicts a mean situation over a certain period of time. Due to the temporal variability an instantaneous current picture is much more complex, and is difficult to interpret. Figure 7.1 is an example of such a picture, showing the instantane-

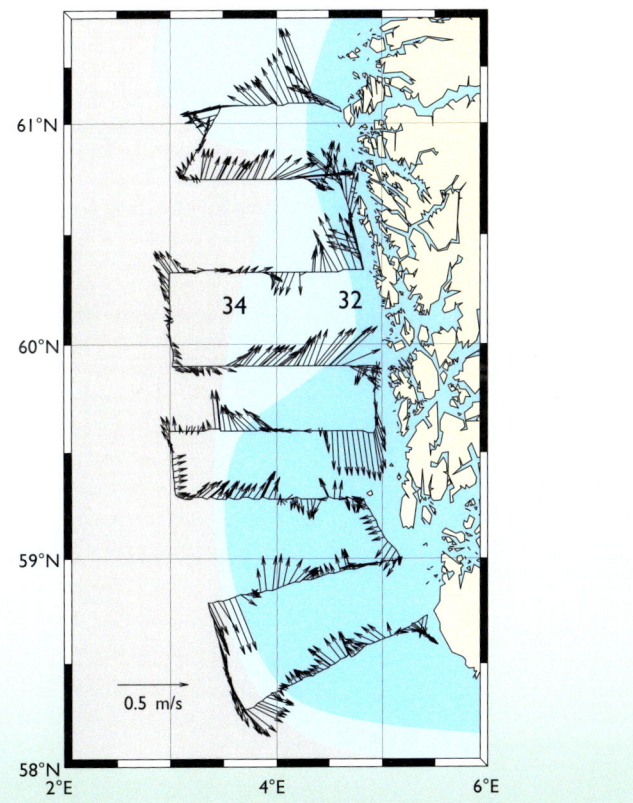

▶ **Figure 7.1** Currents at a depth of 20 m observed by an Acoustic Doppler Current Profiler (ADCP) from a moving vessel, together with the surface salinity isolines 32 and 34, in May 2002 (H. Søiland, IMR, personal communication).

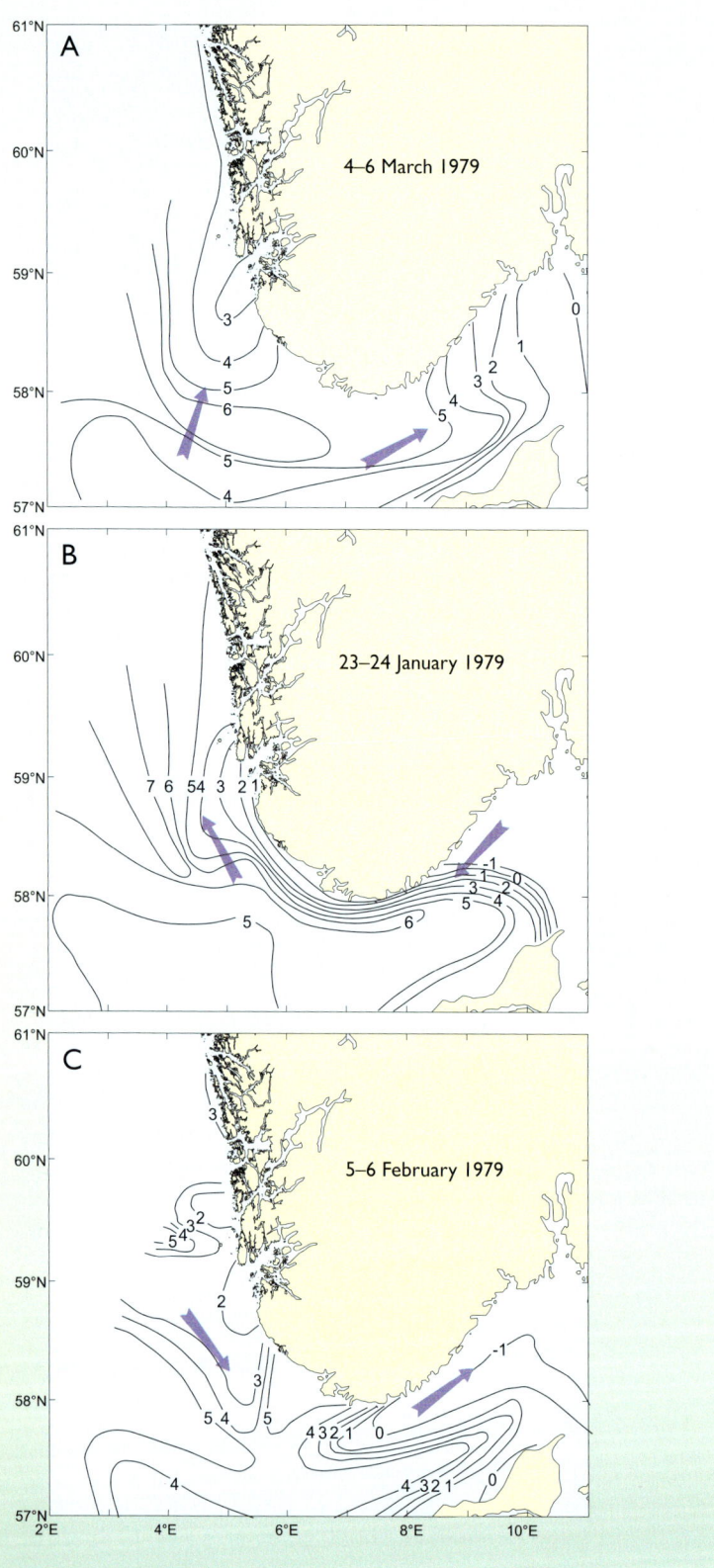

▶ **Figure 7.2**
Sea-surface temperature during the winter of 1979 (Aure and Sætre, 1981). Arrows indicate dominant wind direction.
A) Blocking of the Skagerrak outflow.
B) Typical pattern during a Skagerrak outbreak.
C) An upwelling event.

ous current at a depth of 20 m observed from a moving vessel by an Acoustic Doppler Current Profiler (ADCP) during a few days in May 2002. The 32 and 34 isohalines at the surface are inserted. A minor tidal current element is included in the observations but this does not fully explain the spatial current variability observed. Even in areas where we expect to find the core of the north-flowing Norwegian Coastal Current, such as north of Karmøy at around 59°30', we see a complete reversal of the coastal current. In order to explain a current picture such as we see in Figure 7.1, we need to include short-term features, such as meandering of the frontal zone between the coastal and the North Sea water, formation of eddies or blocking and outbreaks of water from the Skagerrak.

7.2 Intermittent outflow from the Skagerrak

The outflow from Skagerrak of low-salinity water of Baltic origin is not a continuous flow but alternates between blocking and outbreak regimes, giving the outflow an intermittent character. This process is controlled by the prevailing winds (Aure and Sætre, 1981). Westerly or southwesterly winds over the area retard or block the outflow and generate a frontal zone across the Skagerrak (Figure 7.2 A). The cyclonic or counter-clockwise surface circulation of the area is then intensified and low-salinity water of Baltic origin covers most of the Skagerrak. When the westerly wind stress is reduced or the wind turns more easterly, outbreaks of Skagerrak water occur (Figure 7.2 B). Southwesterly winds are probably the most effective blocking agent and blocking seems to be a necessary condition for the outbreaks to occur. The easterly winds that usually follow a blocking period, will intensify and affect the duration of the outbreaks. These are mostly confined to the upper 50 m and to within 20 km of the coast. The most powerful outbreaks seem to occur when easterly winds in the Skagerrak occur in combination with high air pressure over the Baltic.

The speed of the Skagerrak water front along the southwestern coast of Norway during an outbreak is generally around 30 cm/s as far north as Utsira and about 20 cm/s further north. The instantaneous volume transport during a typical outbreak event is around $0.5 \; 10^6 \, m^3/s$ or 0.5 Sverdrup. Total transport, however, is a function of the duration and this may vary from days up to months. Most of the out-flowing water seems to be drawn from the Baltic. During an outflow period that lasted from 7 to 30 January 1937, with strong easterly winds over the Skagerrak, there was a fall of 110 cm in the sea level of the Baltic (Aure and Sætre, 1981).

There appears to be a maximum in the frequency of Skagerrak water outbreaks as well as in the Baltic outflow in February and August (Aure and Sætre, 1981). During outbreaks in extreme cold winters water of sub-zero temperatures has approached the region around Bergen. The influence of the outbreaks tends to decrease northwards along the coast. The low-salinity water that propagates along the coast during an outbreak also penetrates into the fjords. In addition to wind-induced density variations at the mouth of the fjord, some of the major inflows to the fjords are caused by outbreaks of water from the Skagerrak. Penetration of cold, low-salinity water into the fjords during the winter has frequently been reported, and in some cases the Skagerrak water has filled the upper 50 m of some of the largest fjords.

Inflows to the fjords of warm and low salinity water in the upper 100 m during the late summer and autumn seem to be major recurrent events. It is likely that these are a result of increased outbreaks from the Skagerrak in August (Aure and Sætre, 1981).

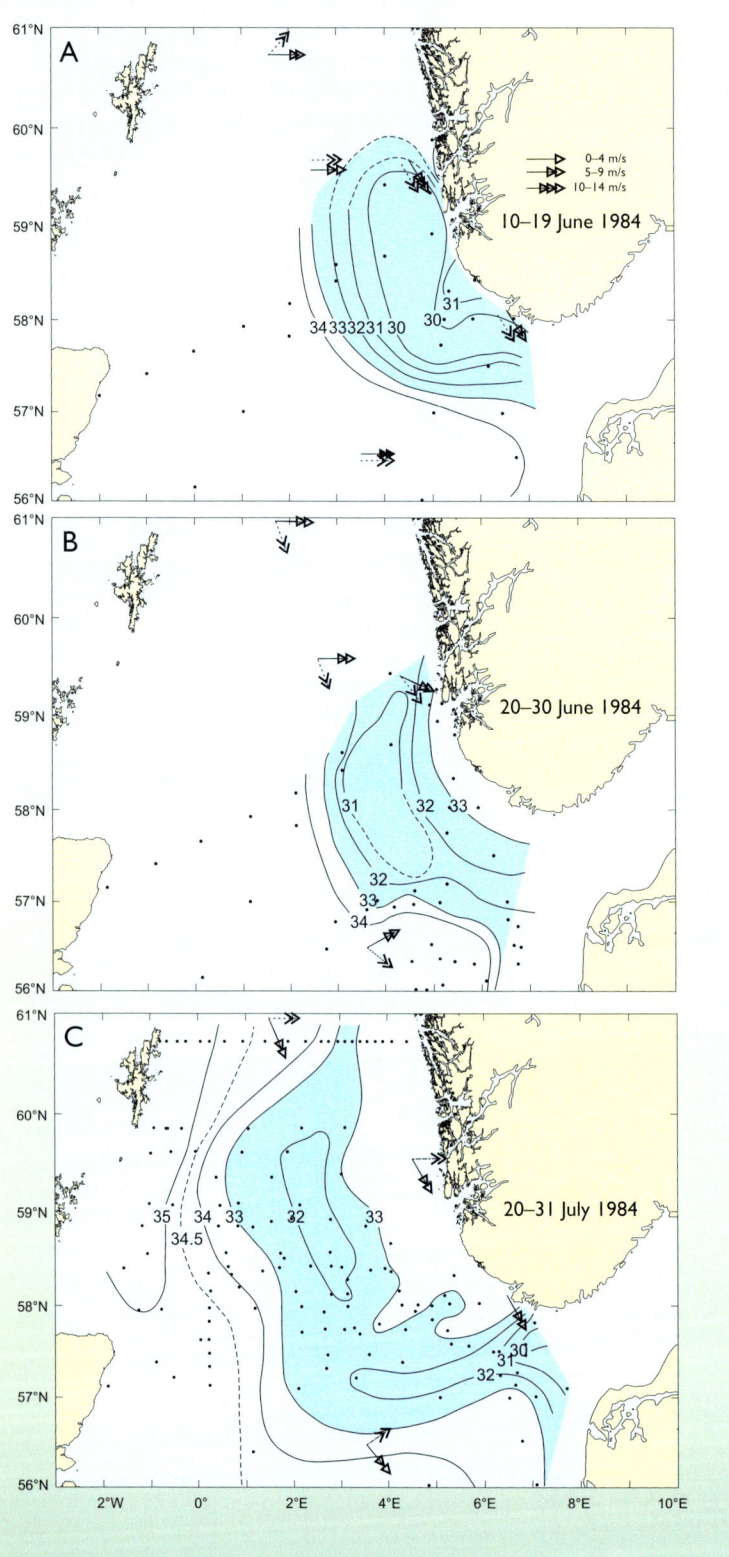

▶ Figure 7.3
Distribution of surface salinities and dominant winds. Unbroken arrows indicate winds during the first five days and dotted arrows the winds during the last five days (Sætre et al., 1988).
A) 10–19 June 1984.
B) 20–30 June 1984.
C) 20–31 July 1984.

7.3 Wind effects, upwelling and horizontal displacement of the Norwegian Coastal Current

The fluctuating wind has a direct and rapid effect on the hydrographic structure of the surface layer. This effect is more pronounced during the summer, when the strong stratification of the upper layer concentrates the wind energy within a rather shallow layer. The wind can move the surface layer in different directions according to the wind and thereby establish zones of convergence and divergence.

Upwelling (Box 7.1) is frequently observed off the southwestern coast of Norway. Figure 7.2 C shows the situation in February 1979, when northwesterly winds off the west coast of Norway forced the Skagerrak outflow seawards. Upwelling close to the coast follows this feature. At the stretch Lindesnes–Jæren upwelling seems to occur in all wind directions between north and southwest,

BOX 7.1 WIND-INDUCED COASTAL UPWELLING AND DOWNWELLING

The wind exerts a force on the ocean surface that is proportional to the square of the wind speed that is setting the surface water in motion. This motion extends to a depth of about 100 m in what is called the Ekman layer, which was named after the Swedish oceanographer V. W. Ekman, who developed the theory of wind-induced currents. Within the Ekman layer the average water particle moves at an angle of 90° to the right of the wind direction in the Northern Hemisphere and to its left in the Southern Hemisphere, a phenomenon known as *Ekman transport*. Ekman transport varies with the wind from place to place, forming zones of divergence and convergence in the surface layer. A region of convergence forces surface water downward in a process called *downwelling* while a region of divergence draws water from below into the surface layer in a process called *upwelling*.

Coastal upwelling and downwelling occur when the wind blows parallel to a coastline, such as the Norwegian coast. If looking into the direction of the wind, you have the land on your left side, the process of upwelling will occur (Figure A). The surface water is driven away from the coast by Ekman transport and is replaced by colder, more saline water from below. If the wind is from the opposite direction and you have land on your right side when looking in the direction of the wind, Ekman transport will be towards the shore, resulting in downwelling and water piling up on the coast (Figure B).

Coastal upwelling and downwelling are important processes in the exchange of water between the coast and the fjords. During upwelling events the surface water of the fjords may be flushed into the open ocean. Downwelling events increase the density of the water at the sill depth and increase the probability of renewal of the deep or bottom water of the fjords.

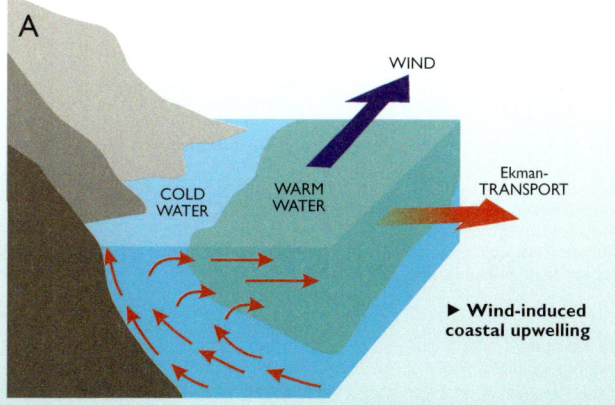
▶ Wind-induced coastal upwelling

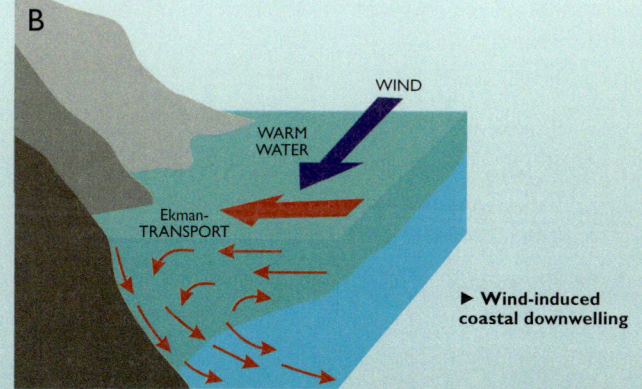
▶ Wind-induced coastal downwelling

► Figure 7.4
Infrared image from the NOAA-7 satellite on 10 June 1984, showing an upwelling situation with colder water along the western coast and outflow of warmer water from the fjords (Sætre et al., 1988).

making this area the location of the most frequent upwelling events anywhere on the Norwegian coast.

Looking at individual years may further clarify the relationship between northerly winds and the westward displacement of the Norwegian Coastal Current during the summer. The summer of 1984 was characterized by the strongest and most persistent northerly winds off southwestern Norway since 1965. The wind distribution was mainly northeasterly until around 10 June, where after the dominant winds were northwesterly. The oceanographic features associated with such wind situations were therefore probably more developed during that year (Sætre et al., 1988).

Around 10 May a major Skagerrak outbreak started, transporting high, temperature, low-salinity water along the west coast. From about 10 June onwards, an upwelling situation associated with the onset of the strong northwesterly winds occurred, with a significant decrease in surface temperature and a rise in surface salinity close to the coast (Figure 7.3 A). At the same time the westward displacement of the Norwegian Coastal Current started and the Skagerrak outflow followed a southern and more offshore route (Figure 7.3 B). This situation persisted until at least the end of July (Figure 7.3 C). As can be seen, the low-salinity water of Skagerrak origin then covered a significant part of the northern North Sea. The maximum depth of the low salinity layer, measured as the depth to the 33 isohaline, was about 30 m. Temporary outflows of Skagerrak water along a southern and more offshore route during the summer are quite common. However, the situation, like that in summer 1984, is probably extremely rare because it lasted for such a long period.

The development of the upwelling along the western coast at the beginning of June 1984 can be followed by a series of satellite images from 6 to 10 June.

The upwelling is characterised by a rapid response to the onset of the northwesterly wind. After one or two days of northerly winds, lower temperatures and increased salinity in the coast-near surface layer can be observed. This pattern started in the northern part around 62°N on 7 June and gradually propagated southwards. By 10 June, the whole of the western coast was affected (Figure 7.4). The time needed to displace the coastal current away from the coast is four to five days. The upwelling had an immediate influence in the coast-fjord water exchange, resulting in a rapid flushing of the top layer of the main fjords into the coastal area (Figure 7.4). As can be seen, the plumes of brackish fjord water retain their form for a long distance from the coast as a result of the westward displacement and the subsequent reduction in strength of the coastal current.

7.4 Fronts and frontal processes

All along its route the Norwegian Coastal Current borders other water masses, mostly of Atlantic origin, which have different properties. At certain times and in some areas the transition zone between the Coastal and the Atlantic Waters is very narrow, while in other cases mixing between the two water masses takes place over a broader area. Such transition zones are called fronts or frontal zones and are found all year long on the coast, appearing as salinity or corresponding density fronts or as temperature fronts.

However, when the horizontal temperature gradients are large, their contrasting temperatures can identify the same fronts more easily. The temperature difference between Atlantic and Coastal waters is often quite marked, with typical horizontal temperature gradients of 0.5 °C/km. Since 1979, infrared satellite images have become more readily available, and the frontal zone can now be regularly mapped using this new tool; e.g. Audunson et al. (1981).

Figure 7.5 shows an example of a frontal structure on the Norwegian coast from February 1980, redrawn from satellite images. The temperature gradient across the fronts in the northern parts may exceed 0.7 °C/km. The location of some of the fronts is relatively stationary, as they are related to the bottom topography. Others are more transient and respond markedly to fluctuations in the wind condition by changing their position, their width and horizontal temperature gradient as well as their waveform.

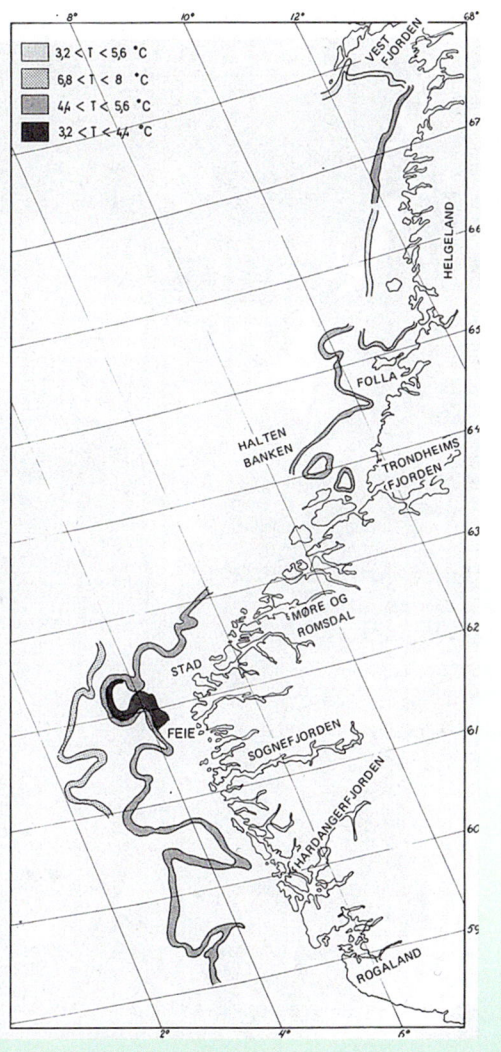

▶ Figure 7.5
Front position and strength on 8 February 1980 (Audunson et al., 1981).

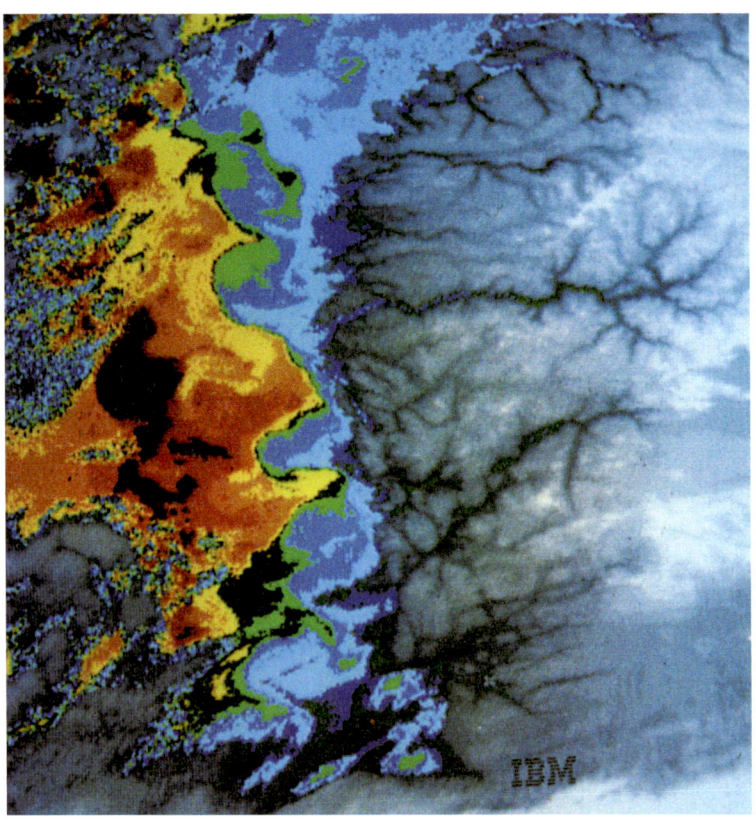

▶ **Figure 7.6**
Infrared image from the TIROS-N satellite on 16 March 1979. Red represents Atlantic water at 6 °C and light blue, coastal water at 3 °C (Anon, 1979).

During the summer of 1969, Mysak and Schott (1977) observed strong fluctuations with periods of 2–3 days in current observations on the continental slope around 62°N, where the upper 50 m consisted mostly of coastal water. These fluctuations varied very little with depth, giving the impression of a barotrophic disturbance. In spite of this, they concluded that the observed current fluctuations can be explained as baroclinic instabilities of the Norwegian Coastal Current.

Johannessen *et al.*, (1996) observed meandering fronts at scales of 10 to 50 km, aligned in the mean northeastward direction of the current on the shelf between 64° and 65°N. This is most likely an effect of topographic steering of the current. There was also a tendency towards concentration of the fronts along the coastal shelf breaks. Evidence of multiple frontal structures was also found. A front is subjected to lateral excursion and wave instabilities, such as the formation of meanders and eddies of 50–100 km wavelength. These processes enahance the mixing between the different water masses at the edge.

Eddies in the Norwegian Coastal Current were first documented during a remote sensing experiment in Norwegian coastal waters in spring 1979 (Anon, 1979). This experiment was a cooperative effort between Norway and the USA and included observations from aircraft and the TIROS-N satellite, as well as from research vessels. Figure 7.6 shows an infrared satellite image off the coast of western Norway from 16 March 1979. Red represents Atlantic Water at 6 °C and light blue represents Coastal Water at 3 °C. The dominant feature is the wavy ocean front between the Coastal and Atlantic Waters. The waves have a wavelength of around 100 km and a wave width of approximately half that length. In the Coastal Current, a double frontal system is quite common (e.g. Figure 7.5). Off southwestern Norway double temperature fronts are frequently observed in winter during a Skagerrak outbreak incident. An inner front then separates mixed coastal water from the colder and fresher water, which has recently been advected into the area from the Skagerrak. This inner front is usually sharper than the ordinary front between the coastal water and Atlantic water.

In spring the main front between the coastal and the Atlantic water may be quite sharp in satellite images that focus on the phytoplankton or the chlorophyll content of the sea. The annual spring bloom starts in the coastal water, which then has a higher concentration of chlorophyll than the Atlantic water. Figure 7.7 shows an example of this.

From 24 February to 6 March 1986 a 10-day field study of the Norwegian Coastal Current off the west coast of Norway was carried out (Johannessen *et al.*, 1989). The study employed infrared images from satellite as well as observations made from a research vessel, and current observations from both drifting and

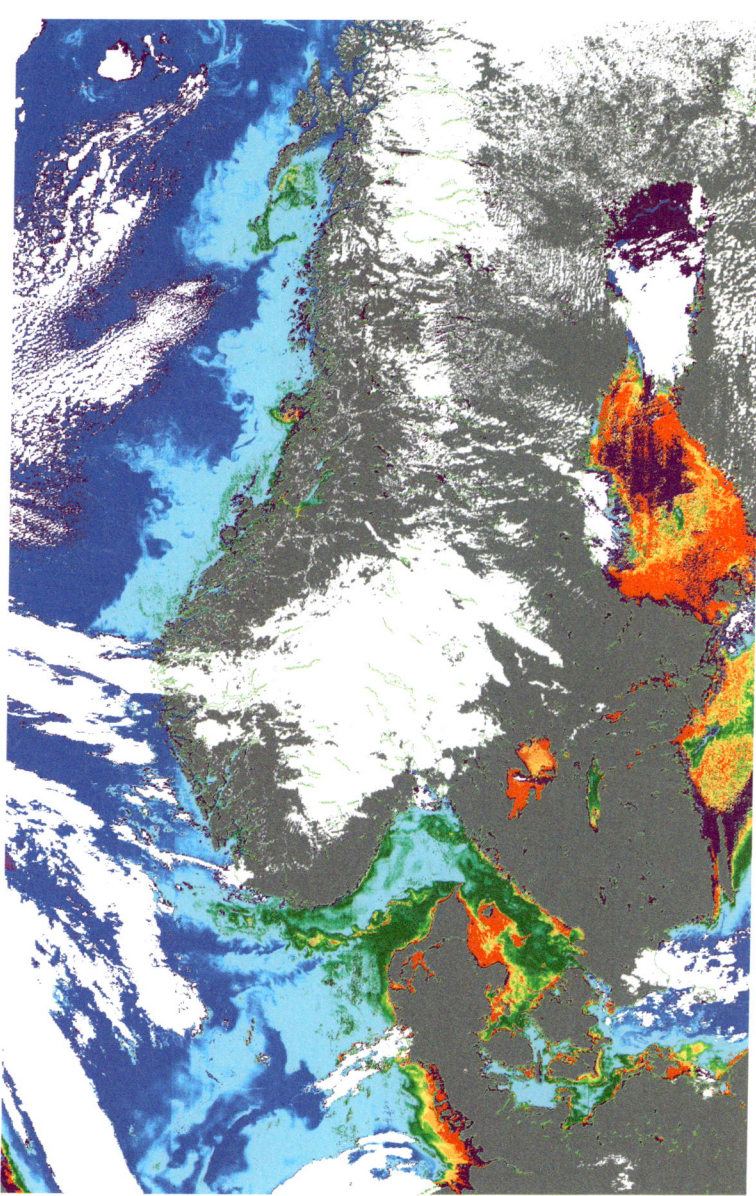

▶ **Figure 7.7** Satellite image from 17 April 2005, showing chlorophyll *a* concentrations off the coast of central Norway.

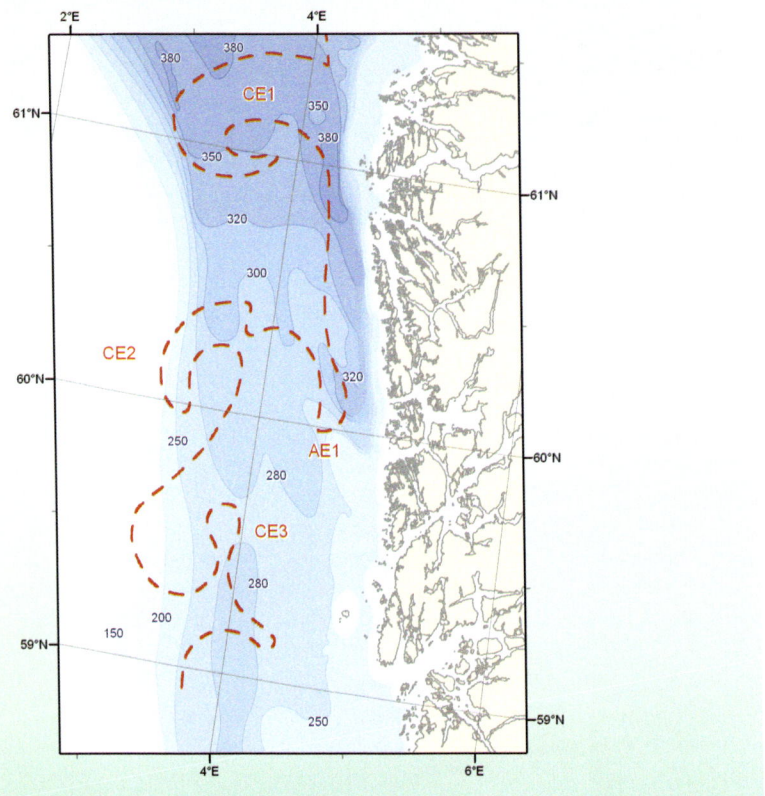

► Figure 7.8
The temperature front between the Coastal and the Atlantic Waters on 25–27 February 1986, as obtained from satellite images, current observations and hydrography. (Redrawn after Johannessen et al., 1989).

moored buoys. Figure 7.8 is a composite figure of the frontal boundary derived from satellite images, current observations and hydrography. Three cyclonic or counter-clockwise eddies, CE 1 to CE 3, and one anticyclonic or clockwise AE 1, can be identified.

The three 40–60 km strongly asymmetric cyclonic eddies extended to the seabed and propagated northwards at a speed of about 5 cm/s. The maximum current speed was reported to exceed 100 cm/s and the velocity was largely constant with depth, confirming that the eddies incorporate a significant barotrophic flow component. Other authors have reported the orbital speed in such eddies associated with the meandering frontal system as reaching nearly 200 cm/s.

Transient eddies may be observed along the whole coast, but most of these are probably generated by baroclinic instabilities and are rather shallow. The physical setting off western Norway appears to be different from that off the rest of the coast. The Atlantic Water along the western side of the Norwegian Trench flowing southwards at a speed of 20–30 cm/s probably plays an important part in the generation of the eddies. Between this inflow and the northward flowing Norwegian Coastal Current there will be a strong velocity shear. During periods of strong outbreaks of Skagerrak water, the shear effect will be amplified. Furthermore, the southward shoaling of the Norwegian Trench tends to give the Atlantic Water an onshore component. The eddies observed in the Norwegian Coastal Current off western Norway can be explained as a combined effect of topographic steering and barotrophic instabilities. However, meanders and eddies generated upstream by baroclinic instabilities may also grow and propagate into the region. Interactions between these types of eddies will take place and complicate the picture.

Characteristic circulation features

Roald Sætre and Jan Aure

8.1 Introduction

The main characteristics of the current conditions along the Norwegian coast can be described by means of a current map. The first attempt to visualise the currents off Norway with the main emphasis on the whole Norwegian Coastal Current was made in 1972 and the most recent one in 1983 (see Chapter 2). New observation methods, both *in-situ* and remote sensing from satellite, have significantly improved our knowledge of the current conditions and their governing mechanisms. The quality of the numerical models used to describe ocean physics have also gradually been improved, especially the spatial resolution and thereby the possibility of including small-scale features.

Ocean currents are the result of a number of periodic and non-periodic movements that cover a wide range of variations in both time and space. In a current map it is therefore impossible to identify the speed and direction of a current by a single vector which is valid for all times. By observing the currents and its variations over a period of time, however, it is possible to establish a mean current pattern. Such mean current pictures should be used with caution, as the short-term current fluctuations often exceed the mean value. As discussed in Chapter 4, the factors that contribute most to current variability off the coast of Norway include wind and atmospheric pressure, tide, topography and freshwater runoff.

The methods used to map the circulation pattern of the Norwegian Coastal Current include:

- Recording current meters suspended from moored buoys to measure the speed and direction of the current at a fixed position (Eularian method).
- Recording the position, i.e. from satellite, of a drifting surface buoy with a drogue suspended at a specific depth (Lagrangian method).
- Continuously observing the vertical current profile from a moving vessel by means of an Acoustic Doppler Current Profiler (ADCP).
- Mapping the horizontal distribution of hydrographic parameters, such as temperature, salinity, nutrients, etc.
- Mapping the horizontal distribution of biological organisms which drift more or less passively with the current, such as plankton, fish eggs and fish larvae.
- Making remote sensing observations of sea surface temperature and ocean colour from satellites.
- Running numerical models in which simplified sets of the hydrodynamic equations of motion are solved by numerical methods in a predetermined grid.

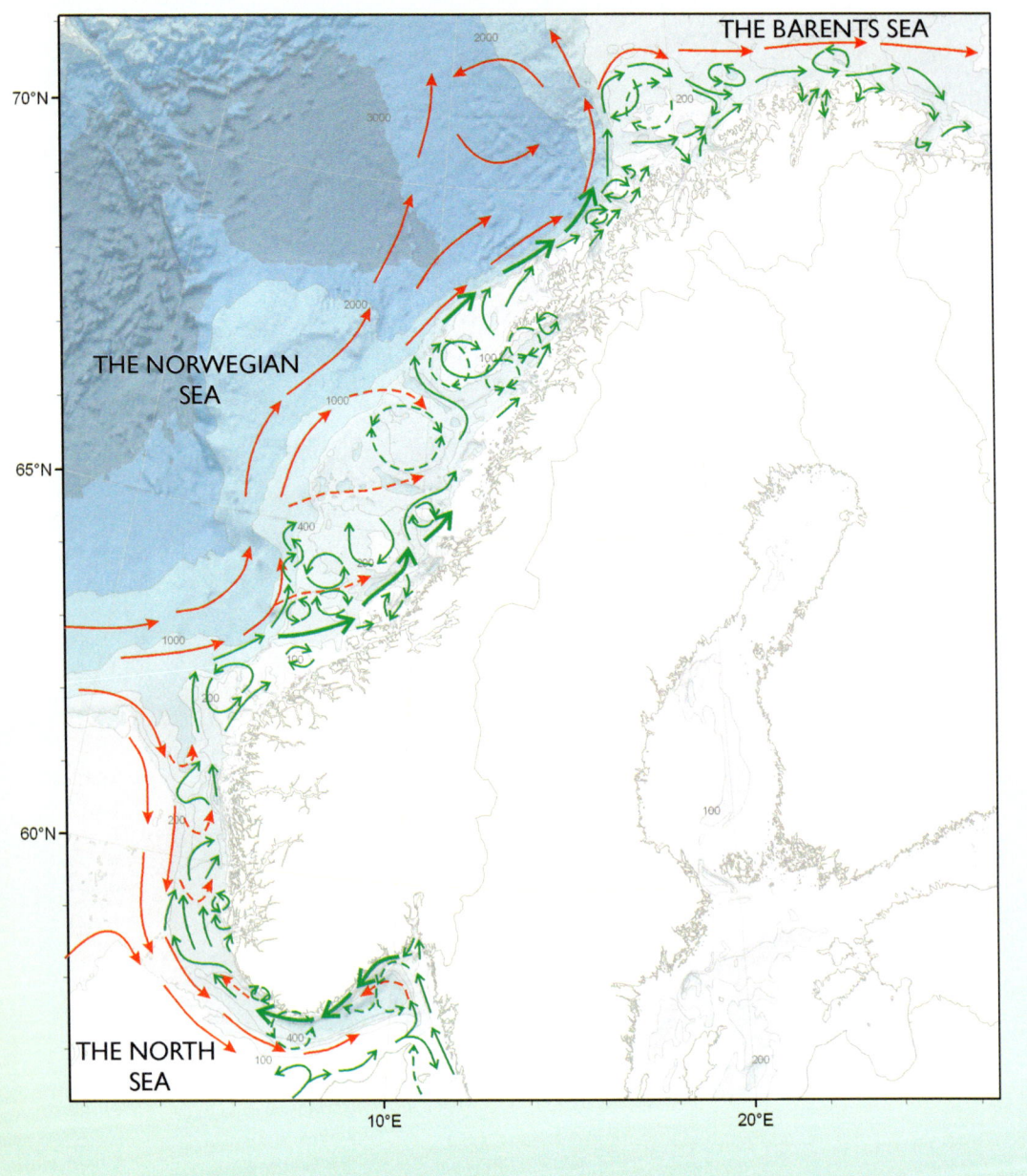

▶ Figure 8.1
Characteristic circulation features of the Norwegian Coastal Current. Red arrows indicate the Atlantic inflow and green arrows the Norwegian Coastal Current, including the tributaries; the Jutland Current and the Baltic outflow. Dotted red lines indicate Atlantic Water flowing below the Coastal Water. Dotted green lines indicate Coastal Current features of transitory or intermittent character.

The current field consists of two parts. One is determined by the internal density distribution, i.e. the temperature and salinity (the baroclinic part), in which the current speed usually decreases with depth (Box 4.2). The other one is due to the inclination of the sea surface (the barotrophic part), and here the current speed is more or less constant from the surface to the bottom Box 4.2).

Sensors in satellites measure the radiation from the earth in the visible and infrared regions of the spectrum, but they need cloud free conditions. Data on the colour and temperature of the sea surface may provide information on water mass types, fronts, eddies etc.

Figure 8.1 is an attempt to bring together all available information on the circulation patterns of the Norwegian Coastal Current into a map. It represents the motion of the upper 50 m of the water column. In such a visualisation it is only possible to include a limited number of important details and a compromise between thoroughness and readability or clarity is always necessary.

8.2 Off southern Norway – south of 62°N

The most conspicuous topographic feature off southern Norway is the Norwegian Trench. All the outflows of water from the North Sea and the Baltic take place along the Norwegian coast. The upper layer of these outflows make up the Norwegian Coastal Current (Figure 8.1). There are three main sources of the water that flows into the Skagerrak: the Jutland Current transports water from the southern North Sea and the German Bight along the west coast of Denmark and into the Skagerrak. There is an inflow of Atlantic Water following approximately the western and southern slope of the Norwegian Trench. Finally, there is an inflow of brackish water from the Baltic. The general circulation pattern of the Skagerrak is cyclonic or counter-clockwise. However, wide temporal variations in this pattern can occur, especially in the upper layer, which are caused by fluctuations in the prevailing winds.

In the shallow area west of Denmark the Jutland Current Water is vertically well mixed due to the strong tidal currents. The changing winds over the area gives the flow of the Jutland Current into the Skagerrak a pulsating character. Occasionally, the inflow may be blocked for several weeks and in such cases the Jutland Current may take a short-cut across the Skagerrak towards the Norwegian coast. In most cases, such blocking events are due to the wind situation, but variations in the Atlantic inflow to the Skagerrak sometimes have a blocking effect on the inflow from the southern North Sea.

The Atlantic inflow follows mainly the 150–200 m depth contour with a current speed of the order of 30 cm/s. During winter this inflow extends from the seabed to the surface, where it may be followed by a band of increased temperature and salinity. During the summer, however, the in-flowing Atlantic water is covered by the Norwegian Coastal Current water.

Figure 8.1 provides a slightly simplified picture of the Skagerrak circulation. The current pattern around Skagen, the northern tip of Denmark, where all the three in-flowing water masses to the Skagerrak interact, sometimes display a rather confusing image. As they cross each other, the water masses cause eddies and meanders in the current. Some more details may be found in Figure 5.9.

Figure 8.2 demonstrates the influence of variable winds on the upper layer circulation. This is results of running a numerical model with 4 km spatial resolution, and four different wind situations are depicted. During strong southwesterly winds (Figure 8.2 A) the outflow from the Skagerrak along the Norwegian coast

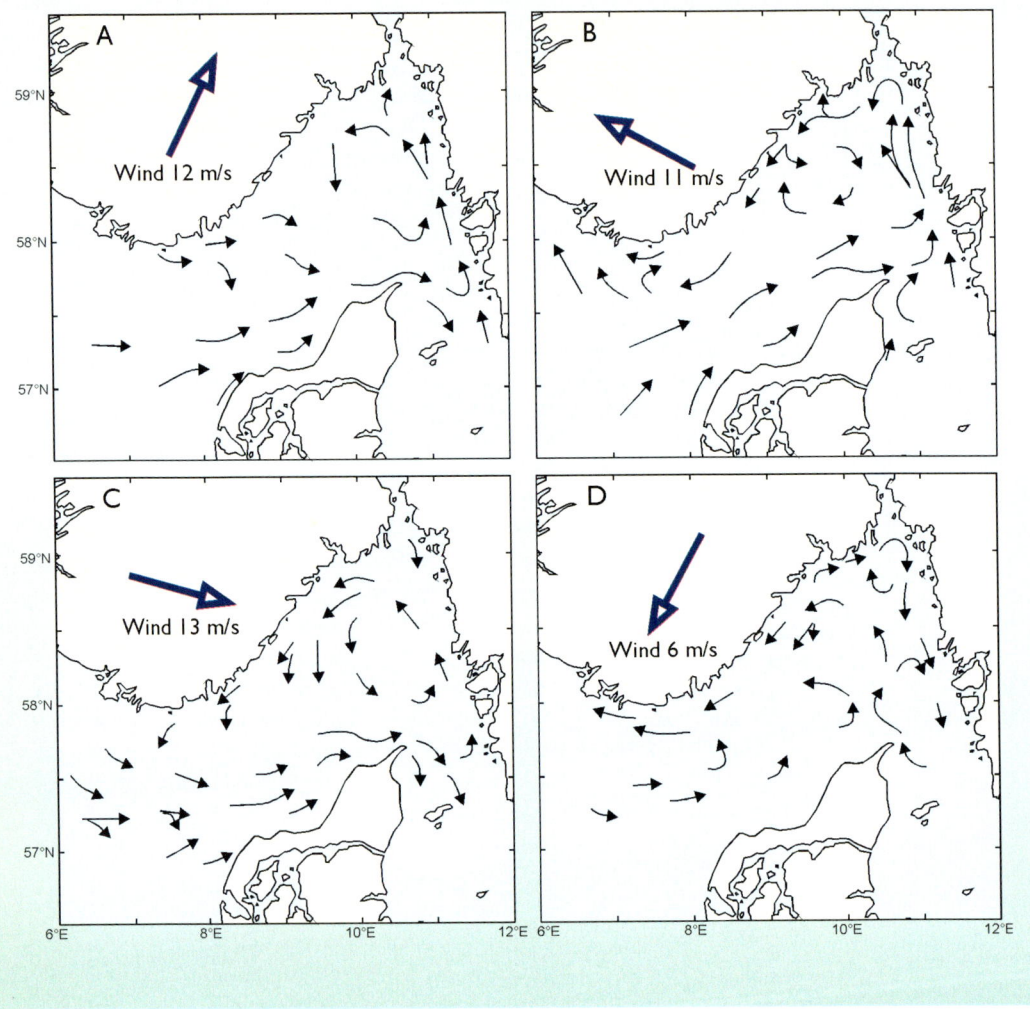

is blocked and the speed of the in-flowing Atlantic Water and Jutland Current rises. Strong inflows of the Jutland Current also occur during periods of southeasterly (Figure 8.2 B) and northwesterly winds (Figure 8.2 C), while they are reduced by northeasterly winds (Figure 8.2 D).

Two large cyclonic eddies are occasionally observed in the Skagerrak, in both numerical models and satellite images; the Langesund eddy in the inner part and the Lindesnes eddy in the outer part of the Skagerrak (Figure 8.1). These features are believed to be governed by special wind situations.

The trajectories of some individual Argos drifters (Box 8.1) are shown in Figure 8.3. These are a clear demonstration of the variability of the currents into and within the Skagerrak. Argos drifter 77 (green marks) with the drogue in 10 m depth was deployed on 1 December in the Jutland Current around 56°N. For three weeks it remained relatively stationary in the area, before it suddenly took off into the Skagerrak at speeds of 50–150 cm/s. In the eastern part it made a full cyclonic turn lasting for about two weeks before it left the Skagerrak with the Norwegian Coastal Current. Argos drifters 107 (blue marks) and 98 (red marks) were both deployed in the Baltic outflow. Their course was more direct, and they

▶ **Figure 8.2**
Currents in the surface layer obtained from a numerical model after one week of winds from different directions (Svendsen, 1995).

BOX 8.1 SATELLITE-TRACKED DRIFTERS (ARGOS DRIFTERS)

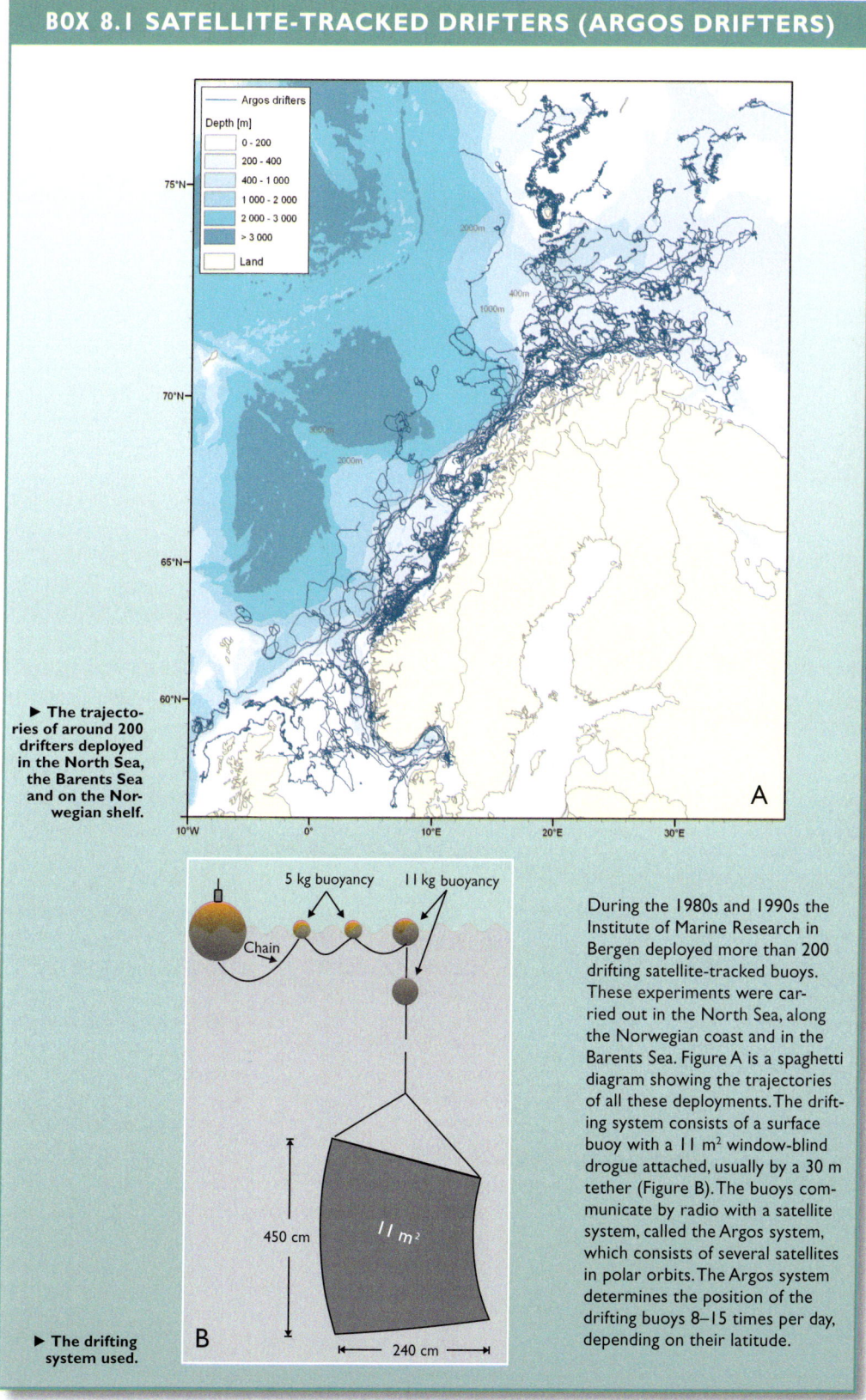

▶ The trajectories of around 200 drifters deployed in the North Sea, the Barents Sea and on the Norwegian shelf.

▶ The drifting system used.

During the 1980s and 1990s the Institute of Marine Research in Bergen deployed more than 200 drifting satellite-tracked buoys. These experiments were carried out in the North Sea, along the Norwegian coast and in the Barents Sea. Figure A is a spaghetti diagram showing the trajectories of all these deployments. The drifting system consists of a surface buoy with a 11 m² window-blind drogue attached, usually by a 30 m tether (Figure B). The buoys communicate by radio with a satellite system, called the Argos system, which consists of several satellites in polar orbits. The Argos system determines the position of the drifting buoys 8–15 times per day, depending on their latitude.

Characteristic circulation features 103

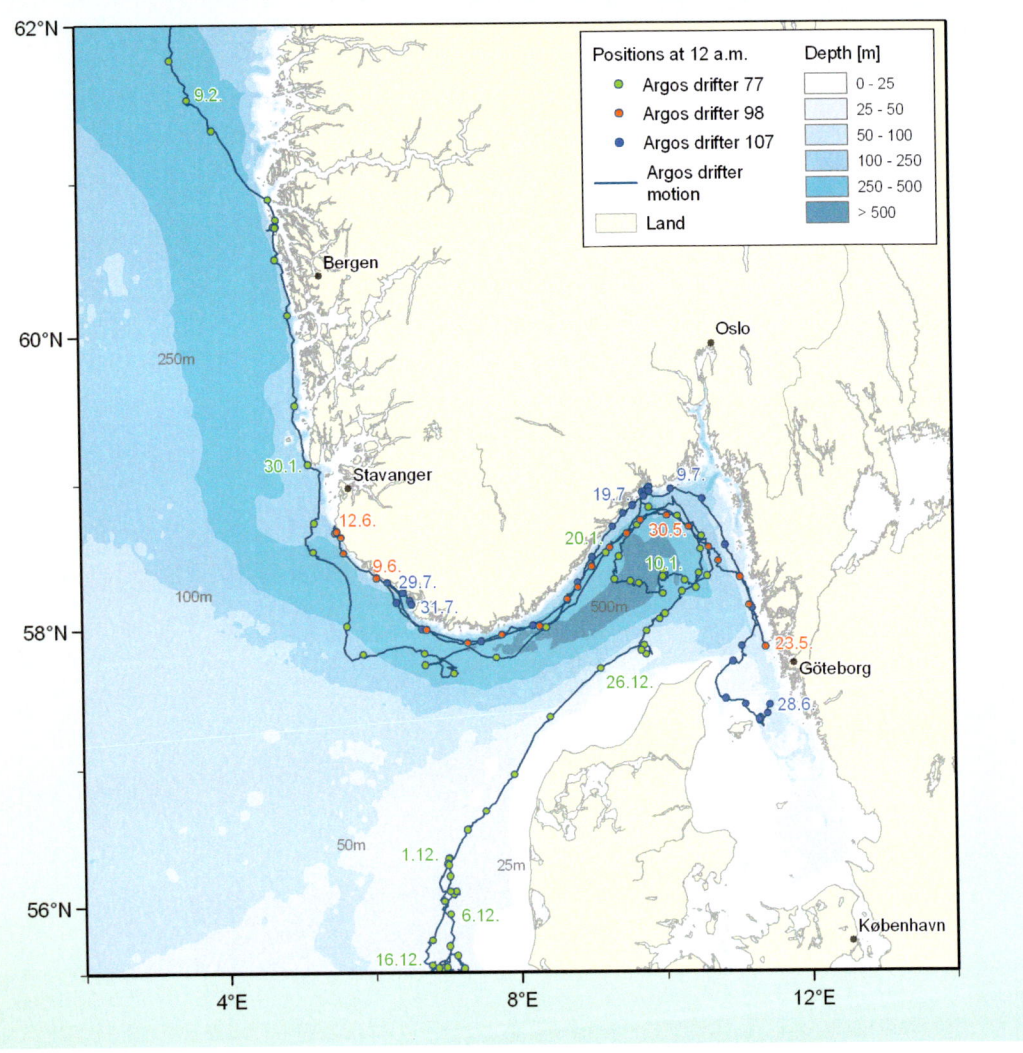

▶ Figure 8.3
Selected individual trajectories of Argos drifters with drogue at depths of 10–20 m.

took 20–30 days to leave the Skagerrak. One of these drifters was caught in the backwater area on the lee side of the headland of Lista. As Figure 8.1 shows, a similar backwater area is located north of the island of Karmøy.

When the Norwegian Coastal Current leaves the Skagerrak its width increases before narrowing again at around 60°N. Farther north at about 62°N it once again becomes wider. These variations in the mean lateral extension of the Coastal Current are mainly an effect of bottom topography. The Norwegian Trench has a sill of 270 m at about 59°N and from here the depths increase both northwards and towards the Skagerrak. The western coast north to around 62°N is characterised by pronounced eddy activity along the frontal zone between the north-flowing Coastal Water and the south-flowing Atlantic Water (see Chapter 7). Sometimes extreme current speeds of up to 200 cm/s are observed within these eddies which are probably associated with sudden outbreaks of water from the Skagerrak, as described in Chapter 7. These eddies typically have a horizontal diameter of 50 km and a depth of about 100 m. The causes of these instabilities by interaction between different water masses may vary, but the strong velocity

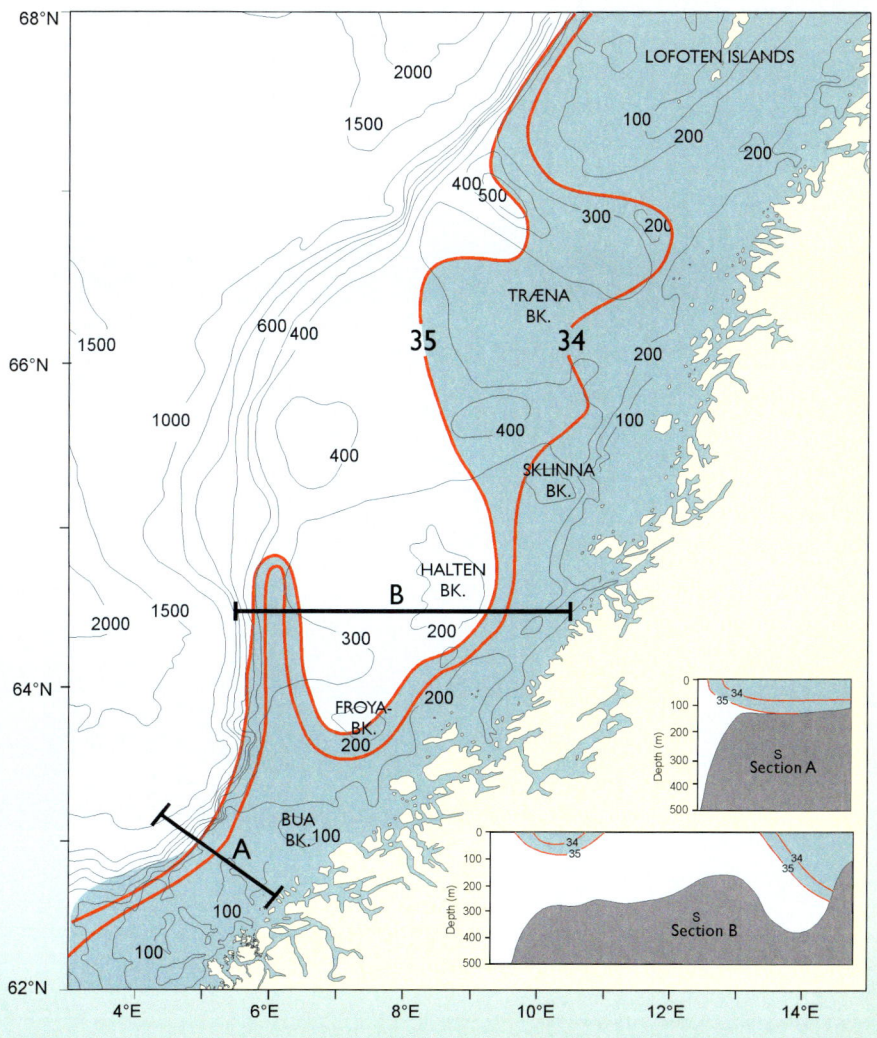

▶ **Figure 8.4**
Bathymetric map with typical water mass distribution for the central Norwegian shelf between 62° and 68°N during winter/spring, indicating surface salinity and the salinity distribution in the two sections A and B.

shear between the two water masses is probably an important factor. The southernward shallowing of the Norwegian Trench favours the re-circulation in the trench of a significant proportion of the Atlantic Water.

8.3 The central Norwegian shelf (62–68°N)

The bottom topography of the region is complex and consists of several shallow banks separated by deeper channels. The typical salinity distribution at the surface and in two vertical sections for the winter/spring period appears in Figure 8.4. The whole southern shelf north to about 63°30'N is rather narrow and the Atlantic Water with a salinity above 35 is restricted to the slope area (Section A). Farther north, the Atlantic Water covers the whole shelf area below 100–150 m, while the upper layer consists mainly of Norwegian Coastal Water (Section B). The salinity distribution (Figure 8.4) further indicates a branching of the Norwegian Coastal Current around 63°30'N. The main branch runs in a narrow band along the coastal side of the Halten Bank, while a secondary branch follows

Characteristic circulation features 105

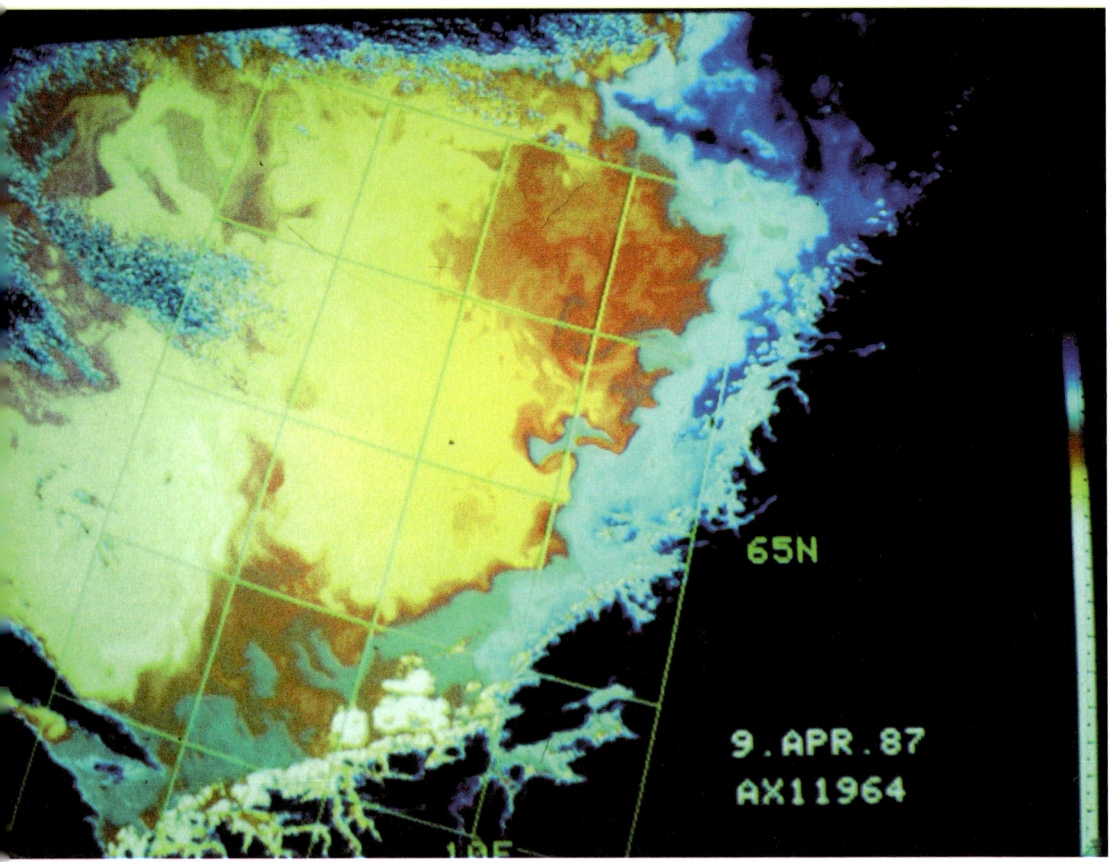

the shelf break northwards on top of the Atlantic Water (Section B). This current branching is also quite clearly seen on infrared satellite images (Figure 8.5).

The following description of the circulation pattern is largely based on the trajectories of around seventy satellite-tracked drifters (Box 8.1) deployed in the area or drifting through it (Sætre, 1999). Figure 8.6 is a trajectory plot of all those that represent the current drift at a depth of 30 m. Circulation on this part of the Norwegian shelf is characterised by the pronounced topographic steering of the current. The current in the Atlantic inflow over the continental slope as well as over most of the shelf is predominantly barotrophic. The inner, principal transport route on the eastern side of the Halten Bank is the Norwegian Coastal Current proper. Between 63° and 66°N it has the character of a coastal jet and its typical wedge-shaped structure is demonstrated in Section B (Figure 8.4). This inner route has higher mean drift speed, lower mean residence time and higher directional stability.

The branching of the Norwegian Coastal Current around 63°30'N brings the Coastal Water northwards along the shelf break. This part of the transport, however, is a minor component. On its way, it mixes with Atlantic Water and gradually loses its identity. It can hardly be traced north of 65°N. Farther north the outer route transports water primarily of Atlantic origin. Atlantic Water, illustrated by dotted arrows in Figure 8.1, also approaches the coast by penetrating into the deeper channels of the shelf.

In addition to the most important drift routes, Figure 8.1 illustrates a number of retainment areas on the shelf. Circulation at these locations includes both

▶ **Figure 8.5**
Infrared satellite image from 9 April 1987, showing the distribution of surface temperature. Blue is cold water and yellow is warm.

cyclonic or counter-clockwise and anticyclonic or clockwise rotation, and the residence time within these areas varies from 10 to 50 days. The anticyclonic retention areas are usually associated with banks such as the Bua Bank, the Frøya Bank and the Halten Bank (Figure 8.4). The banks appear to induce a topographically trapped quasi-stationary eddy or a so-called Taylor column that reaches the surface during the rather vertically homogeneous winter situation. As mentioned above, the whole shelf area is covered by Atlantic Water below 100–150 m. The pronounced topographic steering of the currents is therefore primarily associated with the Atlantic Water, from which the topographic effect is transferred to the Coastal Water above. During the summer, however, with its further development of vertical stratification, there is a decoupling of the upper layer from the water below it, making the upper layer more exposed to the variable winds.

Topographic steering of the current and the possibility of its water being partly trapped in retention areas characterise the inner route. Neither tidal influence nor current meandering of the time scale of some days are conspicuous. The tidal currents are small compared to residual currents and are mainly reflected in the shape of minor pulses in the speed of the current. Changing meteorological conditions do not appear to significantly modify the persistent current pattern along this route.

Figure 8.7 shows the trajectories of Argos drifters 89 (red marks) and 161 (green marks), indicating their daily position. The inner drifter made an anti-

▶ **Figure 8.6**
Trajectory plot of all the drifters deployed or drifting into the shelf between 62° and 68°N. The depth of the drogue was usually 30 m, and most of the drifters were deployed in March–April.

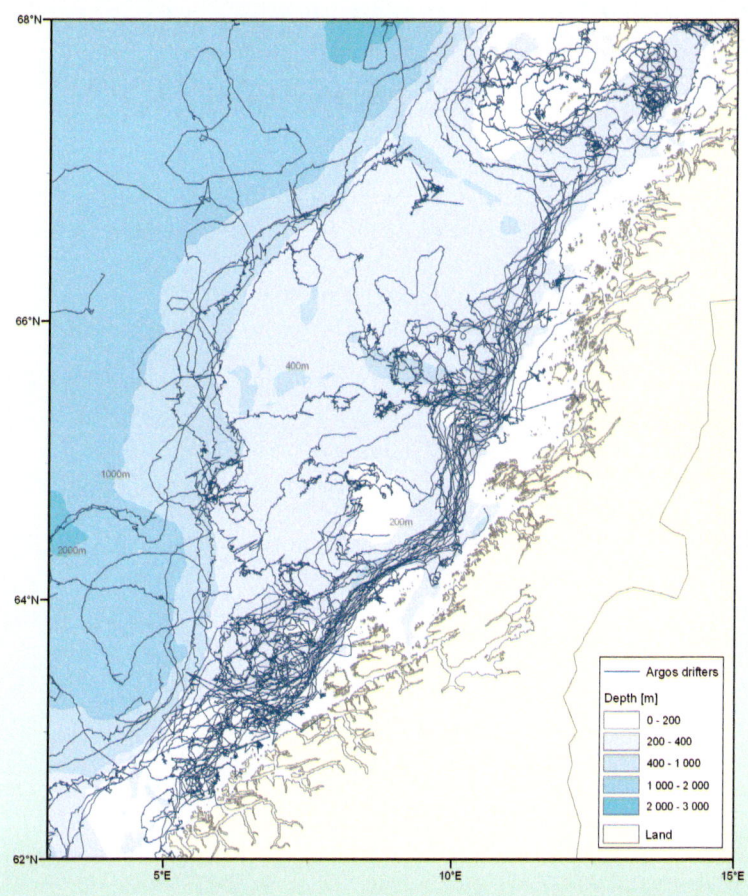

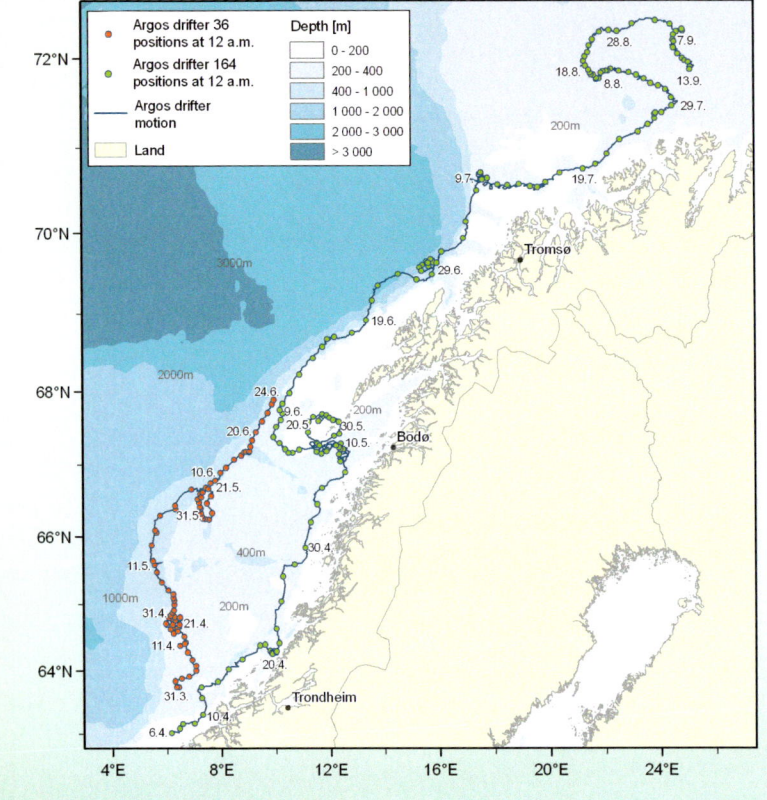

► **Figure 8.7**
Selected individual trajectories of Argos drifters 89 and 161.

cyclonic turn around the Frøya Bank and then was held for about 25 days in the retardation area north of the Sklinna Bank. Later, it appears to be caught in an eddy in the outer part of the Vestfjord between the Lofoten Islands and the mainland. The outer drifter had a maximum drifting speed of about 50 cm/s and made a cyclonic turn during a period of about 25 days. It then returned to the shelf edge and drifted northwards.

Figure 8.8 shows the trajectories of Argos drifters 36 and 164. Drifter 36, following the outer route along the shelf edge, had maximum daily mean speeds of about 40 cm/s. The drift showed two areas of relaxation in the current, in both of which anti-cyclonic movements were dominant during a period of 10–15 days. Argos drifter 164 accelerated from east of the Halten Bank to north of 67°N. Here it was caught by the anti-cyclonic circulation in the retardation area over the shallow bank southeast of the Lofoten Islands, where it spent about 25 days before it continued its journey along the isobaths further north. Usually the highest mean drift speeds (more than 60 cm/s) are found along the shelf break between 69° and 71°N. It passed the Tromsøflaket along the inner route and later moved very slowly around in the Barents Sea east of Tromsøflaket for approximately 45 days.

The mean transport speeds between 63° and 68°N for all the drifters depicted in Figure 8.6 along the inner and the outer route were about 15 cm/s and 10 cm/s, respectively.

Figure 8.9 exemplifies some of the results that were obtained from a numerical model (Vikebø, 2005). The figure shows the monthly mean current at a depth of 20 m in March 1985 on the central Norwegian shelf between 63° and 65°N.

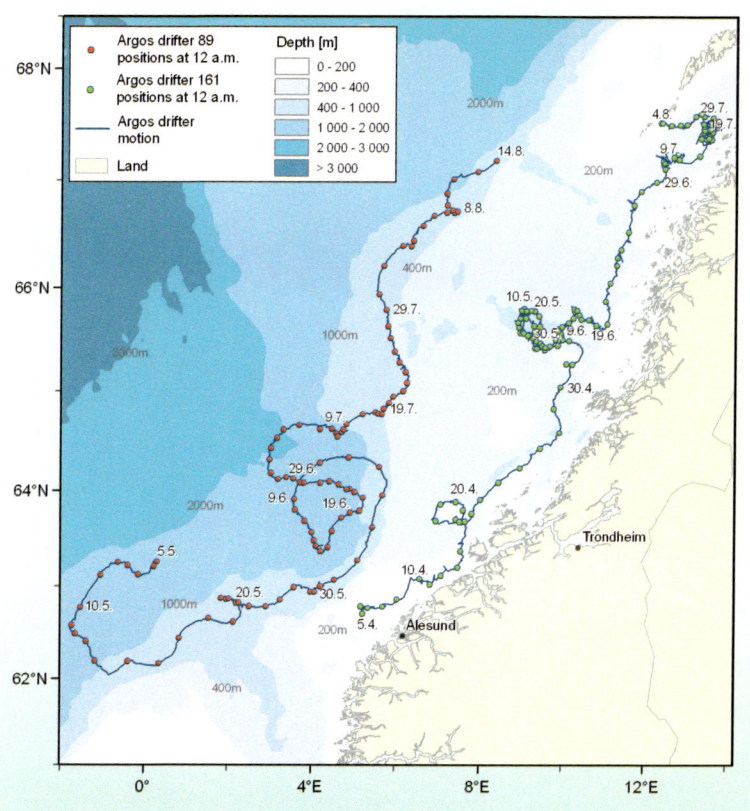

▶ **Figure 8.8**
Selected individual trajectories of Argos drifters 36 and 164.

▶ **Figure 8.9**
Mean current at depth of 20 m between 63° and 65°N in March 1985 as obtained from a numerical model (Frode Vikebø, personal communication).

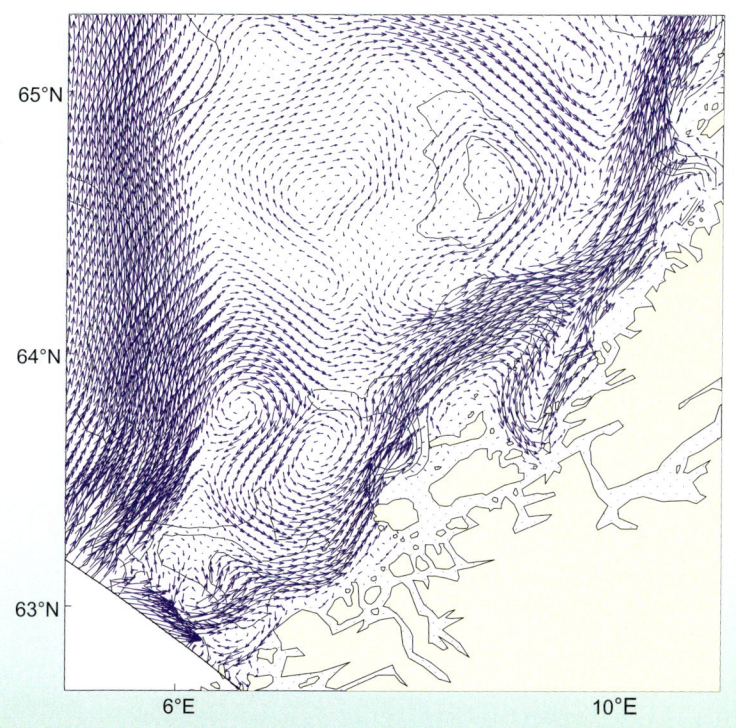

Characteristic circulation features 109

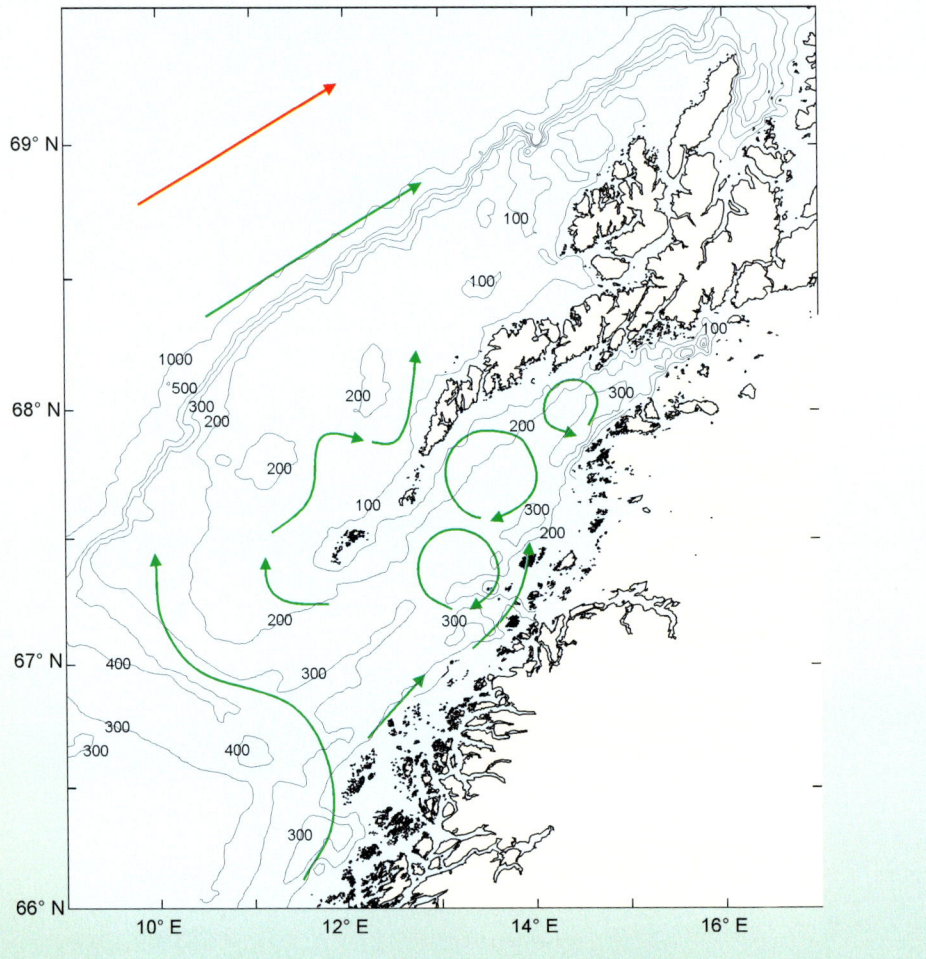

▶ Figure 8.10
Schematic map of the Vestfjord area showing the location of eddy features observed by drifters and from satellite images (Mitchelson-Jacob and Sundby, 2001).

The inner coastal jet is quite conspicuous and eddies of various dimensions, both cyclonic or anti-cyclonic, can be seen.

There is often a distinct temperature front between the Norwegian Coastal Current and the Atlantic Water masses farther offshore. This front tends to be most pronounced during the winter, when horizontal temperature gradients of 0.5° C/km are often observed. The circulation features observed by the drifters along this front can be explained by meanders along the frontal zone. As the directional stability of the current is extremely high over the continental slope, meanders probably only influence the upper 50 m. The outer transport route seems to be characterised by meander-like instabilities of 40–80 km horizontal scale and a 10–25 days time scale. These meanders propagate northwards at a speed of about 10 km/day. Some meanders were pinched off seawards, forming anti-cyclonic eddies.

In recent years, the main spawning areas of the Norwegian spring-spawning herring have been on the central Norwegian shelf between 62° and 68°N. The herring seems to prefer to spawn in areas of topographically induced quasi-stationary eddies, and such behaviour is probably an important recruitment strategy for this stock.

8.4 Off northern Norway – north of 68°N

Between 66° and 67°N most of the Coastal Current makes a left turn and follows the 200–300 m depth contour on the northern side of the Træna Trough, passing to the west of the Lofoten Islands. A small proportion of the coastal water penetrate into the Vestfjord. Eddies of variable size and location appear to be a characteristic feature of the Vestfjord. These were mostly found at three locations (Miechelson-Jacob and Sundby, 2001), as seen in Figure 8.10, and their diameter largely depends on the width of the fjord. There is some evidence that wind conditions are a factor in determining the direction of the rotation of the Vestfjord eddies.

Figure 8.11 shows the mean circulation at a depth of 20 m in March 1985 as derived from a numerical model (Vikebø, 2005). In this case, there is an anticyclonic eddy in the inner part of the Vestfjord, while the circulation in the outer part of the fjord is cyclonic. North of the Vestfjord the Atlantic Water tends to be found off the continental edge, and a typical current speed will be 50 cm/s. In this area the width of the continental shelf is very narrow; off Andenes it is only around 10 km.

The movement of the Coastal Water is strongly influenced by the bottom topography. Figure 8.12 illustrates this for the surface layer at the three minor bank areas Sveinsgrunnen, Malangsgrunnen and Nordvestbanken between 69°30'N and 70°30'N. The circulation map is based on hydrographic data, drifters and the distribution of cod eggs (Sundby, 1984). This situation is probably more representative of winter conditions, while during the summer, with its greater vertical stratification, the topographic signal in the surface layer is less pronounced. As observed on the central Norwegian shelf, an oscillating inflow of Atlantic Water will intrude into the troughs between these banks. There is some

▶ **Figure 8.11**
Mean circulation at depth of 20 m in the Vestfjord area in March 1985 as obtained from a numerical model (Frode Vikebø, personal communication).

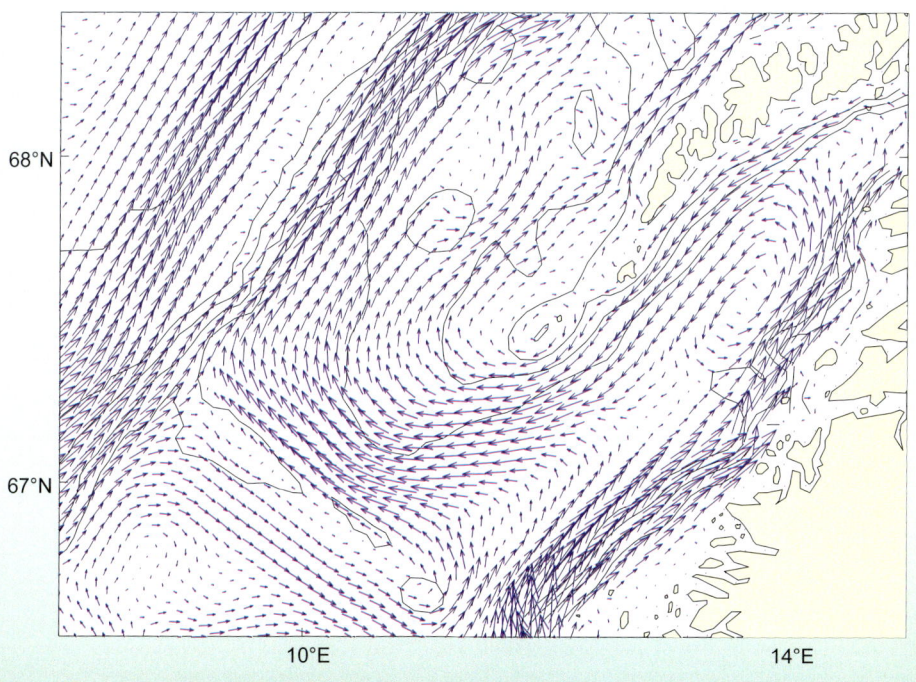

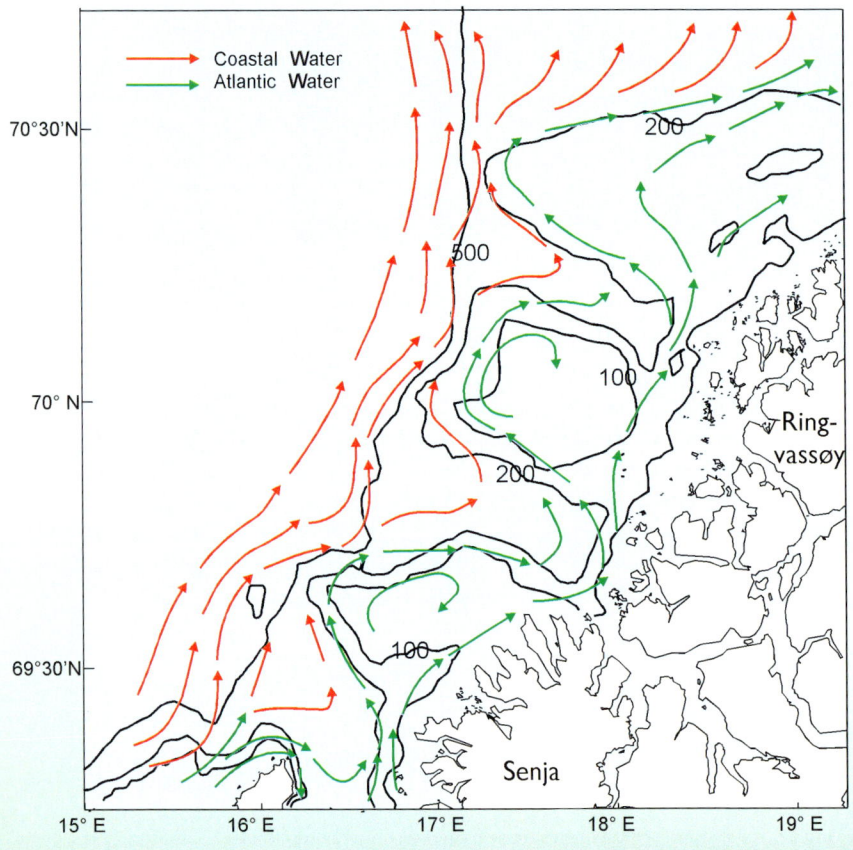

▶ Figure 8.12
The circulation in the upper layer between 69° and 71°N during a typical winter situation.
Green arrows: Coastal Water.
Red arrows: Atlantic Water
(Sundby, 1984).

evidence that this variability mostly reflects wind fluctuations, with wind-driven upwelling induced by the topography.

Farther north, at the Tromsøflaket Bank, the Atlantic Water splits into two branches; one continues northwards to Spitsbergen along the slope of the western Barents Sea while the other turns eastwards into the Barents Sea itself. Coastal water covers a large part of this bank and the circulation is dominated by one or more anti-cyclonic eddies. Tromsøflaket is a well-known retention area with a longer residence time for water, and in spring and early summer for cod larvae that have been transported there by the Norwegian Coastal Current, mainly from the Lofoten area. It appears as though the Coastal Water has two potential routes: one following the depth contour northwards and circling around the Tromsøflaket, while the other takes the short-cut flowing through the trench on the southern edge of the bank as recorded by the drifter shown in Figure 8.7. On the Tromsøflaket proper both the current and its directionality are usually weak. The persistent winds are the key factor that determines whether the major part of the coastal water will follow the northern or the southern route. During northerly winds the Coastal Water is influenced by an offshore Ekmann transport and mainly follows the northern route, thereby increasing the likelihood that fish larvae will be transported northward towards Spitsbergen. Under wind conditions of this sort, the residence time of the water on the banks also increases. When the winds blow from the south, on the other hand, the speed of

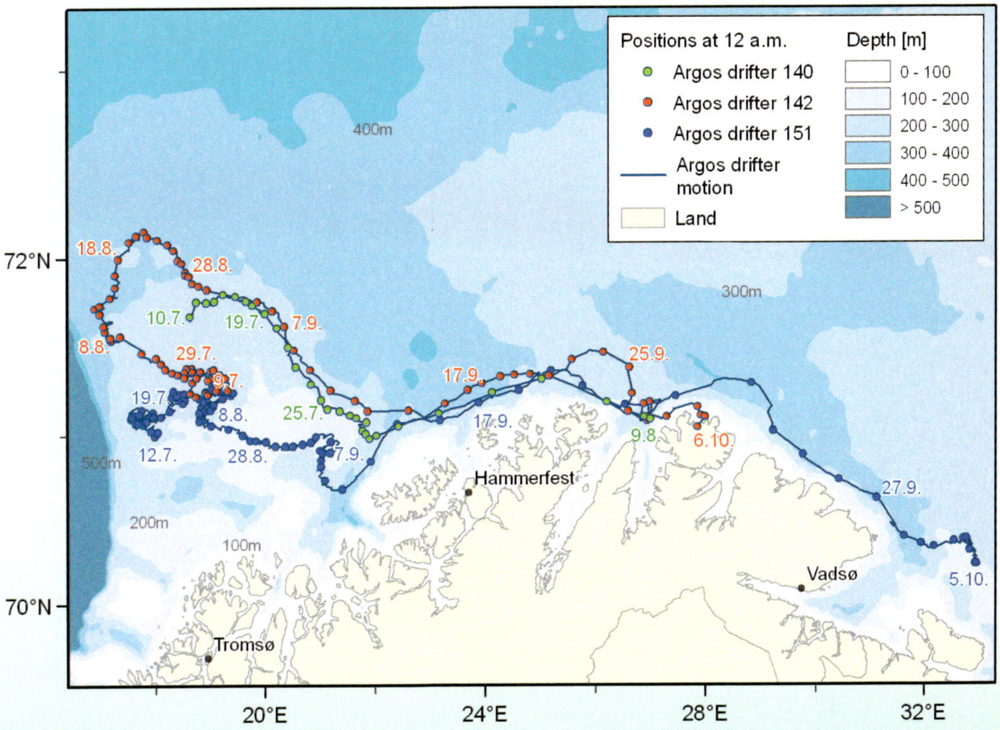

▶ **Figure 8.13**
Some selected trajectories of Argos drifters. The depth of the drogues was usually 30 m.

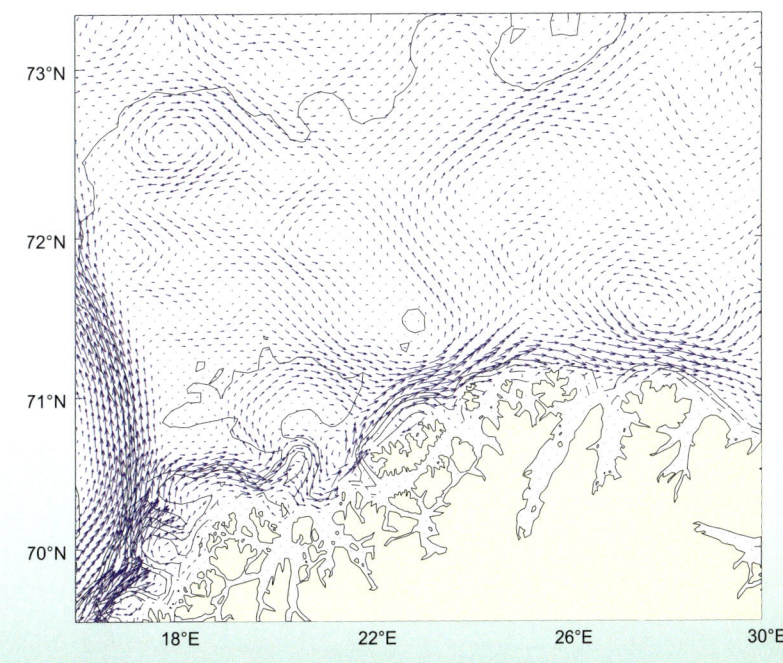

▶ **Figure 8.14**
Mean circulation at depth of 20 m north of 69°30'N in March 1985 as obtained from a numerical model (Frode Vikebø, personal communication).

the coastal water increases and its residence time on the banks is shorter. The water is concentrated closer to the coast and mainly follows the southern route into the Barents Sea.

East of the Tromsøflaket, a characteristic feature is the narrowing of the Coastal Current off the North Cape and an offshore deflection of the current takes place farther to the east. This deflection is probably a result of bottom topography and is often depicted as an empty area in the horizontal distribution of fish larvae transported by the coastal current. Two cyclonic eddies over depressions in the bottom are also often observed.

Figure 8.13 shows the trajectories of Argos drifters 140 (green marks), 142 (red marks) and 151 (blue marks). All the drifters confirm the role of Tromsøflaket as a retention area. Drifters 142 and 151 were trapped on Tromsøflaket for approximately two months and drifter 140 for one month. East of Tromsøflaket, however, all the drifters indicate a rather swift eastward-flowing coastal current. The mean circulation at a depth of 20 m in March 1985 (Figure 8.14), derived from a numerical model (Vikebø, 2005) reflects the main features of the circulation reasonably well. In this case, most of the Coastal Water is following the southern route on the southern side of the Tromsøflaket.

Coast/fjord water exchange

Jan Aure, Lars Asplin and Roald Sætre

9

9.1 Introduction

The coasts of many countries around the world are characterized by the existence of semi-enclosed bodies of water, which stretch inward from the coast. These bodies of water are often termed *"fjords"*. A fjord is a narrow inlet of the sea between cliffs or steep slopes, which is the result of marine inundation of a glacial valley. However, there are great differences between fjords, topographically, climatologically and dynamically (e.g. Gade, 1986). Originating from glacial erosion (see Chapter 3), many fjords have a sill in the mouth area, as well as one or more deep basins (Figure 9.1). The fjords in Norway exist at all different physical scales up to a length of about 200 km and depths of more than 1300 m.

The fjords are a main source of freshwater to the Norwegian Coastal Current. The mean annual freshwater runoff from Norway is around 12 000 m^3/s or about 400 km^3/year and all this flows through the fjords. This contributes to approximately 40 % of the total freshwater transport of the coastal current (see Chapter 4). The southwestern region of Norway contributes most to the total freshwater discharge. The runoff shows a clear seasonal signal, usually with a maximum in May–June and a minimum in February–March. In western and central Norway a distinct rainy autumn is characteristic, producing a secondary runoff maximum during that season.

▶ **Figure 9.1**
Idealised water exchange processes in a typical fjord.

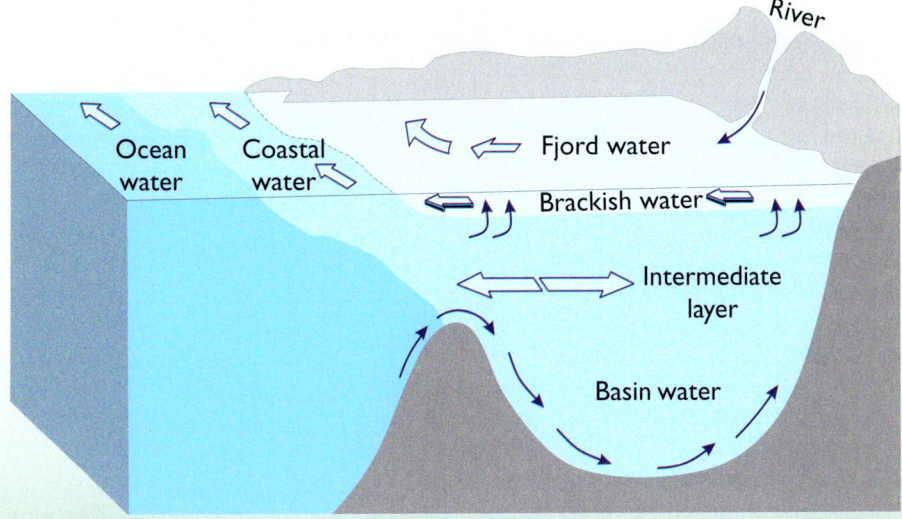

Interannual fluctuations in the runoff may be large. Figure 9.2 shows the annual runoff to the longest Norwegian fjord, the Sognefjord, during 1986–2001, together with the winter index of the NAO or the North Atlantic Oscillation (see Chapter 4.3). A high NAO winter index means strong westerly winds over Northern Europe and increased precipitation; i.e. there exists a strong correlation between the NAO winter index and annual precipitation in Western Norway. As Figure 9.2 shows, the amount of precipitation is a good proxy for the fjord runoff.

The freshwater inside the fjords is forced by pressure out towards the coast, flowing in a brackish upper layer. Several processes inside the fjord, e.g. winds and tides, mix the freshwater with the more saline water beneath it. When the freshwater reaches the coast, it is as a mixed water mass of slightly lower salinity than the coastal water mass. Various mixing processes inside the fjord, as well as the water exchange from the coast to the fjord, determine the difference in salinity between this water mass and the coastal water mass. This difference varies over time, according to the amount of freshwater and the variable strengths of the different mixing agents. The low-salinity water from the fjords may influence both the properties and the dynamics of the Coastal Current.

The effect of the rotation of the Earth, the Coriolis force (see Box 4.1), is important in fjords wider than the baroclinic Rossby radius of deformation (Box 9.1) or around 2–3 km. In such "broad" fjords the circulation is deviated to the right by the Earth's rotation, and thus the strongest flows often occur nearest the land. The influence of the rotation of the Earth in a "broad" fjord also depends on the vertical stratification of the fjord and the depths of the upper layer (Box 9.1). These conditions display seasonal variations and in consequence, the effect of the of the rotation of the Earth on the circulation varies in the course of the year. In narrow fjords, however, the effect of the Earth's rotation is insignificant and the water transport is relatively constant across the fjord, i.e. the inflow and outflow of the upper layer take place at different depths, and the fjord circulation can be regarded as a two-dimensional process.

The water masses of the fjords are stratified, and the dynamics involved are complex. The bottom topography will significantly influence the flow. The physical conditions of the fjords are primarily reflecting the properties of the Coastal Water. The fjord's influence on the Coastal Water is mainly through the

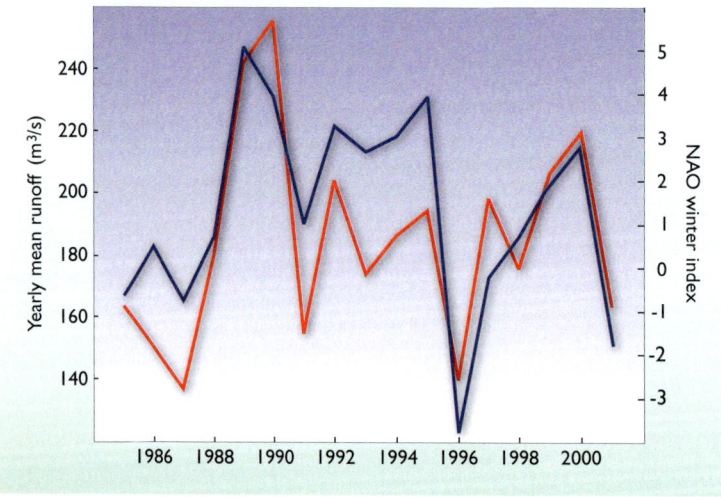

► Figure 9.2
Yearly mean runoff to the Sognefjord and the winter North Atlantic Oscillation (NAO) index.

freshwater runoff, which on an annual basis is close to the volume of the freshwater outflow from the Baltic. The brackish outflow from the fjords, which have a strong seasonal variation, has an effect on the vertical stability of the coastal water and thereby on its heating and cooling. It largely determines the Rossby radius of deformation and thus influences the dynamic features of the Coastal Current. This chapter considers only coast/fjord water exchange, and a complete description of the physical condition of the fjords is beyond its scope.

9.2 Exchange processes

Water exchange between fjord and coast is the sum of different flow components, which are described in more detail below. Usually the flow is linear, i.e. the total flow is the sum of all components. It is useful to characterize exchange in terms of depth zones: surface layer, the intermediate layer (between surface layer and

BOX 9.1 ROSSBY RADIUS OF DEFORMATION

For an earthbound observer every object moving freely on the surface of the earth appears to curve slightly from its initial path – to the right in the northern hemisphere and to the left in the southern hemisphere. This is an effect due to the rotation of the Earth – the Coriolis effect (see Box 4.1). *The Rossby radius of deformation* gives us the length scale at which the geotrophic balance between the Coriolis force and the pressure gradient force (see Box 4.2) becomes important. On a non-rotational earth a flow of water would be straight down the pressure gradient. In a geotrophic flow, however, the flow is *deformed* by the Coriolis force and the flow is parallel to the isobars (lines of constant pressure), rather than crossing them. When an ocean basin or the width of a fjord is much larger than the radius of deformation, the currents or the circulation of that ocean or fjord will be in geotrophic balance.

The radius of deformation also gives us the scale of feature such as fronts (distance across the front) and the width of ocean currents. These will be "close to" the radius of deformation, which in oceanography means something like from half to twice that radius.

This term is also used in connection with coastal upwelling (see Box 7.1) and can be demonstrated by a two-layer model (see Figure). An upper layer of thick-

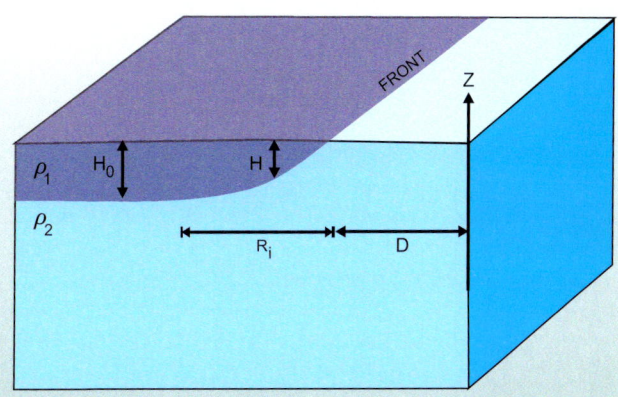

ness H_o and density ρ_1 overlies a deep layer of density ρ_2. Imagine a wind blowing along the coast with land on its left side. Due to the Coriolis force, an offshore Ekman transport take place in the upper layer, which has to be compensated for by upwelling of water along the coast. After some time the interface between the two layers reaches the surface and the front starts to move offshore. The distance, D, it has moved offshore, is proportional to the strength of the wind. The width of the region within which the interface rises to the sea surface, R_i, is the *internal Rossby radius of deformation*.

In a two-layered ocean the equation for the Rossby radius of deformation, R, can be expressed as:

H_o is the depth of the upper layer, ρ_1 and ρ_2 the density of the upper and lower layer respectively and f the Coriolis parameter. The radius of deformation depends heavily on the stratification or $(\rho_2-\rho_1)$, the thickness of the upper layer, H_o and the Coriolis parameter. As this parameter increases with latitude, R will be reduced on higher latitudes. Seasonal fluctuations in both H_o and $(\rho_2-\rho_1)$ in the fjords and coastal waters result in similar variability in R. In Norwegian fjords and coastal waters, the internal radius of deformation is about 3–5 km.

$$R = \frac{(g' H_o)^{1/2}}{f} \quad \text{where} \quad g' = \frac{g(\rho_2-\rho_1)}{\rho_2}$$

sill depth) and the deep layer below sill depth (Figure 9.1). The surface layer transports fresh water from the fjord to the coast. The flow in the intermediate layer is governed by differences in the internal pressure between coast and fjord and transports are considerably greater than in the surface layer. The deep layer contains the basin water, which tends to be stagnant, and is only occasionally exchanged with the coastal water.

Freshwater-induced exchange
This so-called *estuarine circulation* occurs in the upper layer of the fjord and is a result of the freshwater runoff. It is driven by the difference in density between the brackish layer in the fjord and the coastal water outside. The seaward flowing brackish water, with typical speeds of 10–20 cm/s, is continuously mixed with seawater and the brackish water transport reaching the coast is usually between five and ten times the freshwater runoff (Gade, 1986). As the annual runoff from Norway is around 12,000 m^3/s, this means that the maximum freshwater-induced coast/fjord water exchange for the whole Norwegian coast is probably considerably below 100,000 m^3/s or 0.1 Sverdrup. Beneath this outflowing surface brackish water there is an inflow of coastal water, which closes the estuarine circulation cycle. The freshwater-driven coast/fjord water exchange is small compared to other exchange components.

Locally wind-driven exchange
Wind stress imposes a mechanical drag on the surface water of a fjord, pushing it in the wind direction and slightly to the right if the fjord is wider than the radius of deformation (Box 9.1). The wind-driven flow decreases rapidly below the surface, and by a depth of 10–20 m, no more sign of this flow is found. Surface speeds more than 100 cm/s may be reached, but the duration of such speeds is usually limited to a few hours. The local wind-driven flow is thus more important as a mixing agent in the surface brackish layer inside the fjord, while its contribution to coast/fjord water exchange is insignificant.

Exchange driven by coast/fjord sea level differences
The level of the sea surface of the coastal water varies due to tides and meteorological forces such as wind and atmospheric pressure. Winds blowing with the land on its right side results in piling up of water and thereby increases the sea level along the coast. In combination with low atmospheric pressure such piling up of water may occasionally reach 1–2 m above normal sea levels, and such cases are called *storm surges* (Box 10.1). However, such extreme sea level variations are rather rare and though coast/fjord water exchange may be strong during these sometimes dramatic events (Box 10.1), its influence on an annual scale is rather small.

Tidal variations dominated by the semidiurnal tide have a much greater influence on the coast/fjord water exchange due to their rapid fluctuations and the relatively large tidal differences. The mean difference in sea level between high and low tide increases from about 0.3 m on the Skagerrak coast to around 2.7 m on the northernmost coast of Norway. This suggests that the theoretical mean coast/fjord water exchange per unit area of the fjord during a tidal period can vary from 6–7 m^3/s per km^2 on the Skagerrak coast to around 60 m^3/s per km^2 on the northernmost parts of the coast (Figures 9.3 and 9.4). Water exchange between coast and fjord due to tidal movements is thus approximately ten times as high on the northern coasts of Norway as on the Skagerrak coast. As the tidal current is driven by horizontal sea level differences, the driving force is the same

BOX 9.2 CAN A FISH STOCK IMPOVERISH ITS ENVIRONMENT?

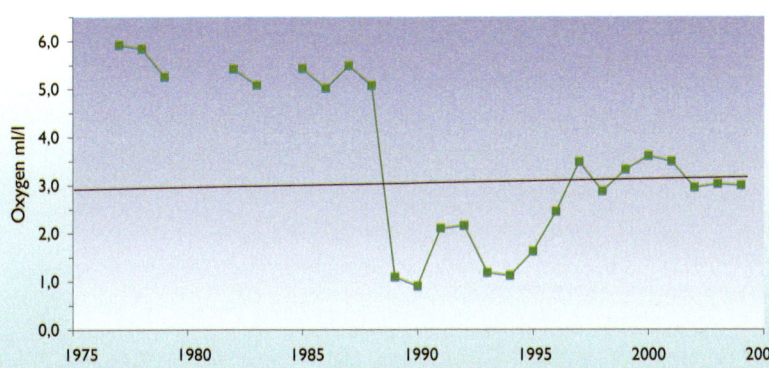

▶ Oxygen (ml/l) at a depth of 200 m in the Ofotfjord outside Narvik in December 1977–2004.

The scientific literature demonstrates clearly that fluctuations in environmental conditions are capable of influencing the migration, growth and recruitment of a fish stock. The following is an example of the opposite effect, whereby a fish stock clearly affects its environment in a negative way.

After the summer feeding season, Norwegian spring-spawning herring migrate to the wintering areas where the stock is stationary until the spawning migration starts in January. The stock collapsed in the late 1960s and the remaining stock wintered in several fjords on the coast. In 1983 the first strong year-class since the collapse was produced and the stock began to recover. Since the winter of 1987–1988 the adult stock has wintered in the Ofotfjord and Tysfjord, which are two side fjords to the Vestfjord close to the Lofoten archipelago. The wintering stock has a biomass of several million tonnes, and such a large concentration of biomass within a relatively small area can be expected to produce environmental changes (Dommasnes et al., 1994).

During the wintering period the herring do not feed, although they do make diurnal vertical migrations. During the hours of daylight the herring are found at depths of 150–200 m and at night they tend to remain at depths of 50–100 m. Figure A shows the oxygen content at 200 m in the Ofotfjord outside Narvik in December in 1977–2004. Until 1988 the oxygen content was around 5 ml/l, but in 1989, after the herring stock entered the fjord, it was drastically reduced to around 1 ml/l. The low oxygen values in the deeper layers of the fjord have persisted. The lowest values are usually observed in January when the herring stock starts its migration out of the fjord.

In recent years, a growing proportion of the herring stock has been wintering outside the Ofotfjord. The lower concentrations of herring produced a marked increase in the oxygen content of the water column in 1995–1997, since when, and until 2004, there has been a relatively stable oxygen content of about 3 ml/l in the Ofotfjord. This suggests that the wintering stock of herring has been fairly constant since 1997.

from the surface to the bottom layers. This means that the tides are capable of establishing flows in the lower layers which usually do not take part in either the estuarine or the local wind-driven circulation.

However, the effective water exchange due to tidal forces is a great deal less than these theoretical values might suggest. The outflowing and inflowing waters mix, and the effective tidal water exchange are usually estimated to be around 50 % of the theoretical value. In fjords with long narrow entrances, the tidal water exchange is further reduced by friction or "choking". However, in such cases, the vertical turbulent mixing above the sill depth increases. The choking effect increases with the tidal difference, and this type of fjord is thus more often to be found on the northern than the southern coast.

Exchange driven by coast/fjord horizontal density differences

Density fluctuations in the coastal water generate horizontal coast/fjord pressure differences, which induce flows in or out of the fjord. The density distribution of the coastal waters off the fjord mouth may be changed either by

- advection of new water masses of different properties
- wind-induced coastal upwelling or downwelling (Box 7.1)

Increased coast/fjord water exchange due to the arrival of new water masses have been observed in southwestern Norway during sudden outbreaks of Skagerrak water (see Chapter 7.2). During extremely cold winters, such as in the early 1940s (Box 10.2), the coastal water masses were largely replaced by water of lower temperature which had been formed outside the area. An inflow to the fjords followed the arrival of these new high-density water masses.

Wind-induced upwelling or downwelling generates density fluctuations in coastal waters that induce coast/fjord pressure differences that are associated with in- or outflowing currents. On the western coast of Norway, persistent southerly winds transport surface water closer to the coast, resulting in downwelling of less dense water. Northerly winds have the opposite effect, as they transport surface water away from the coast, resulting in upwelling of denser water (Box 7.1). Along the western Norwegian coast the coastal current associated with upwelling or downwelling typically extends about 5 km offshore, a distance comparable to the internal Rossby radius of deformation (Box 9.1). The transport is rapid, and during an upwelling event about half of the upper layer of the fjord may be replaced in the course of one or two days (Asplin et al., 1999).

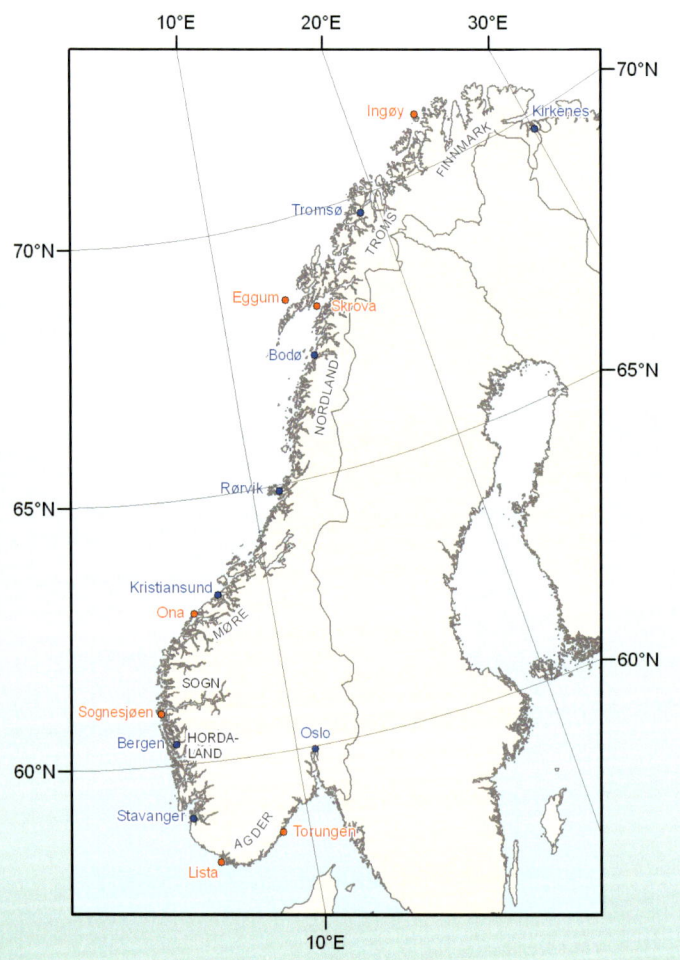

▶ **Figure 9.3**
Locations of the observation points referred to in this chapter.

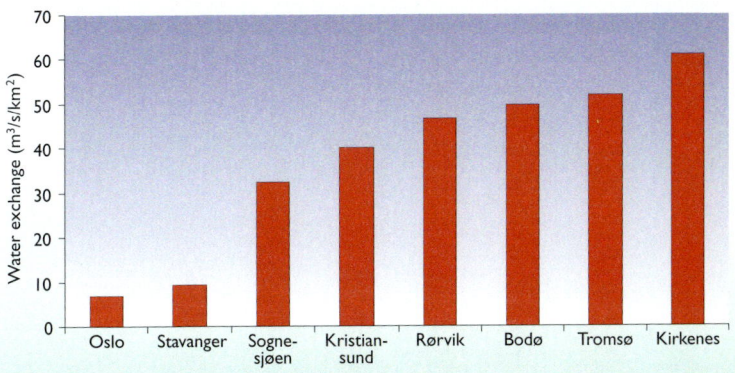

▶ **Figure 9.4**
Mean theoretical water exchange (m³/s) per unit fjord area (km²) driven by the mean tidal difference at selected positions on the coast (Figure 9.3).

Figure 7.4 illustrates such an upwelling situation in June 1984, when the whole upper layer of the fjords of western Norway was flushed to the coastal area in the course of about a week.

The average result of downwelling is an outflow to the coast in the lower part above sill level and a compensatory inflow in the upper part. During upwelling situations this type of circulation is reversed. The residence time of the intermediate water masses ranges from a few months to as much as approximately half a year. The mean coast/fjord intermediate water exchange has been shown to be proportional to the amplitude fluctuations or standard deviation of the variations in coastal water density, integrated from the surface to the sill depth (Aure et al., 1996). On the basis of observations from fixed oceanographic stations (Figure 9.3), the standard deviation expressed in water mass per unit area has been calculated for sill depths ranging from 10 to 100 m during a mean year (Figure 9.5). The figure demonstrates that the standard deviation, and thus intermediate water exchange, increases with increasing sill depth and decreases from south to north along the coast. This is a consequence of the fact that the vertical stability of the coastal water masses is reduced as we move north. There are also clear seasonal variations, with the strongest intermediate water exchange taking place during the summer.

The coast/fjord water exchange during an upwelling or downwelling event takes place during a remarkably brief period (about a day), and the absolute value of the volume of the involved flows is 10–100 times as great as the volume fluxes associated with the classical estuarine circulation (Stigebrandt, 1990).

In fjords that are wider than the baroclinic Rossby radius of deformation (Box 9.1), the movement will be influenced by the rotation of the earth; the Coriolis force. We are all familiar with waves on the sea surface. Similar waves can also be generated in deeper layers, such as at the boundary surface between two water masses of different density. Such waves are called internal waves and may have an amplitude of about 30 m without any elevation of the sea surface. They occur regularly in stratified water where horizontal density differences exist. These differences may arise from coastal upwelling or downwelling driven by regional winds. In the region of the fjord mouth, upwelling and downwelling are prevented, and after approximately 24 hours a coastal-trapped internal wave will be generated (Asplin et al., 1999). This wave will start to propagate at a speed of about 50 cm/s, with the land on its right-hand side (Figure 9.6).

This wave propagates until mixing and dissipation by forces or complex topography break it down, but it typically lasts from several hours to a few

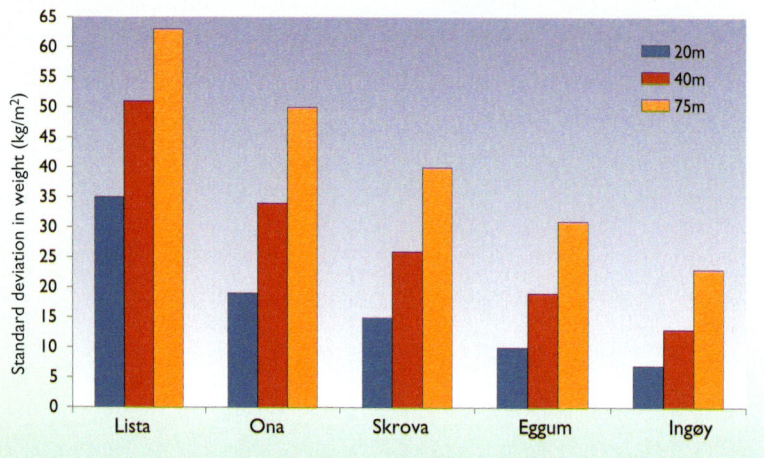

► Figure 9.5
Mean annual standard deviation in the mass (kg/m²) of the water column from the surface to a given depth along the coast.

days. The wavelength is long, and the associated currents appear in the form of a coastal current (Figure 9.6). The wave is typically a two-layer wave, with a much deeper lower than upper layer. The flow in the lower layer is therefore often negligible in comparison with other components of fjord flow. The depth of the upper layer is typically 10–30 m, and waves created by downwelling tend to be deeper than those created by upwelling. This means that the coast/fjord water exchange flux during downwelling is greater than during an upwelling event. The flow in the upper layer is normally into the fjord for the downwelling case and out of the fjord for the upwelling case. The current speed is typically 20–30 cm/s, and the current extends down to 20–30 m below the surface during downwelling and 10–20 m during upwelling events. The width of the current in the upper layer approaches the internal radius of deformation, i.e. 3–5 km. Associated volume transports in the upper layer are of order 10^4 m³/s, which means that it is about ten times as large as the volume flux related to the estuarine circulation. The net water exchange with the Coastal Water from this process is uncertain, although the sum of many such internal waves is part of what being characterized as intermediate water exchange driven by horizontal density differences.

An example of this process is from the results of a numerical fjord model as illustrated in Figure 9.7. The area concerned is just north of Bergen and the grid size for the model is 500 m. The wind is southwesterly and the wind speed 12 m/s. After about 24 hours several internal coastal trapped waves have developed and started to penetrate the fjords. After 48 hours the waves have reached the head of Fensfjorden. A current of about 20 cm/s in the upper 30 m from the coast to the inner parts of Fensfjorden is then established, as is a weaker current in opposite direction below this depth.

Let us now consider a typical fjord with the following characteristics: entrance width: 1000 m, sill depth: 60 m and surface area: 80 km². Figure 9.8 shows the contribution to the total coast/fjord exchange from the intermediate water exchange driven by horizontal density differences and the tidal water exchange in such a fjord in different sectors of the coast (Figure 9.3). On the Skagerrak coast the intermediate water exchange contributes about 90 % of the total, while the tidal influence is insignificant. However, the importance of this latter component increases as we move north, and on the northernmost coast, the density-driven water exchange is only around 30 % of the total. From western Norway and northwards, the total coast/fjord water exchange is approximately

▶ **Figure 9.6**
Propagation of coastal water into a fjord by a coastal-trapped wave.

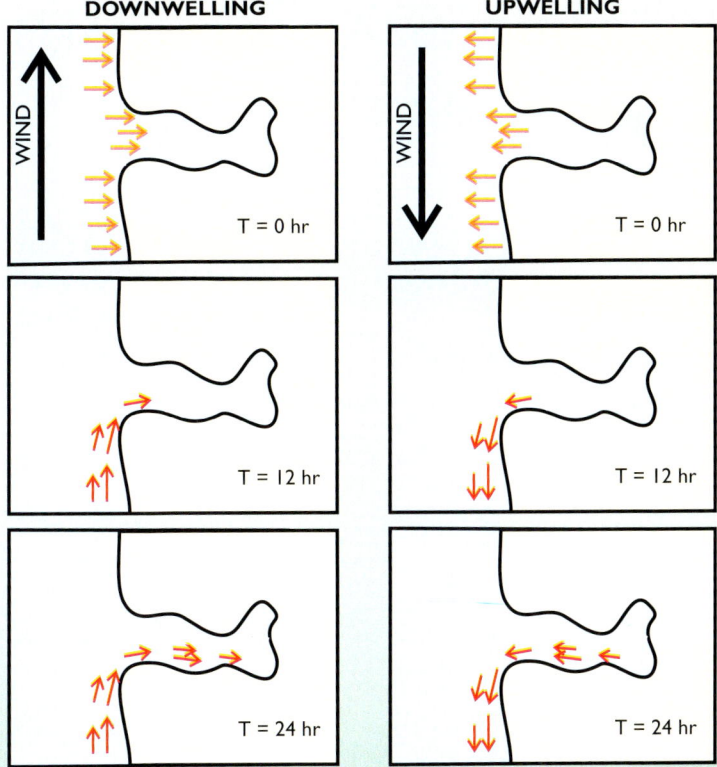

▶ **Figure 9.7**
Current direction and speed (red colour is the highest speed) after 24 and 48 hours at 10 m depth from a numerical simulation in a fjord area north of Bergen during southwesterly wind of 12 m/s along the coast. Maximum current speeds are 20–30 cm/s and the distance between the arrows is 1 km.

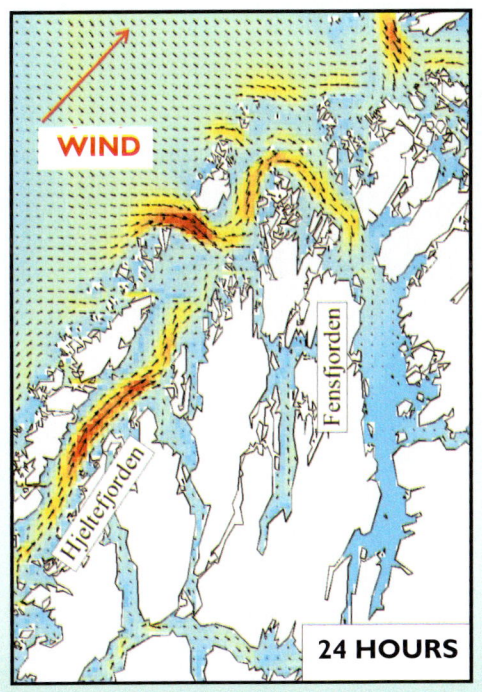

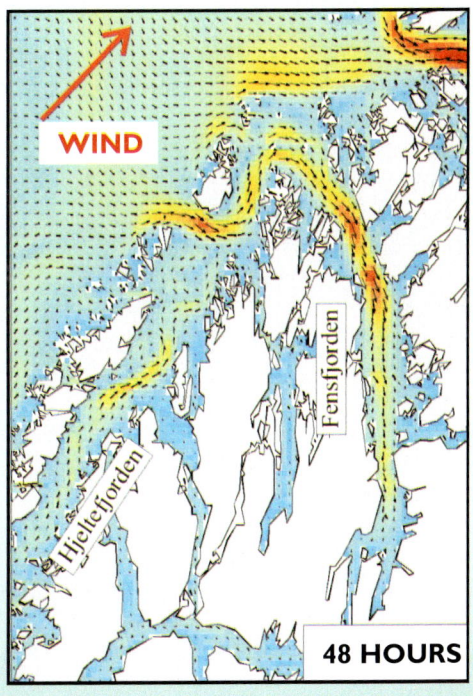

Coast/fjord water exchange 123

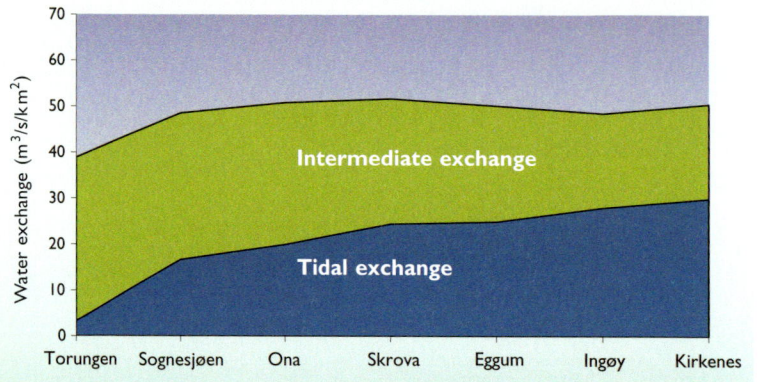

▶ **Figure 9.8**
Calculated effective tidally induced water exchange (m³/s) and intermediate water exchange (m³/s) per unit area (km²) of a specific fjord.

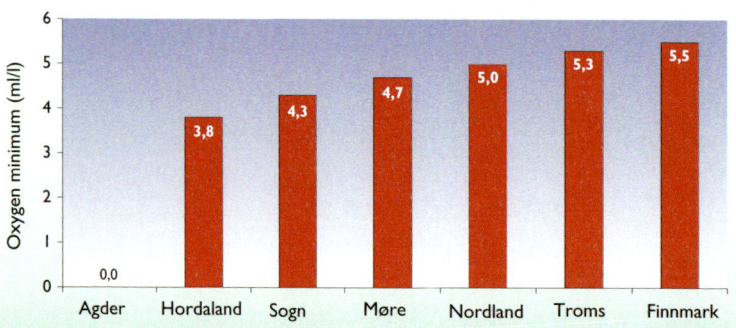

▶ **Figure 9.9**
Calculated minimum oxygen level in basin water before it is renewed.

the same as the increasing tidal influence roughly compensates for the reduction in the density-driven water exchange.

Renewal of the basin water
Tidal forces are also important for the vertical turbulent mixing of fjord basin water (Figure 9.1). Greater tidal amplitudes increase vertical mixing, thereby increasing the rate of reduction in the density of the basin water. This in turn leads to a higher probability of renewal of the basin water and thus to improved oxygenation. In coastal areas with relatively small variations in density, the frequency of inflows to the basin water will increase. The decrease in fluctuations in the density of the coastal water and the rise in the tidal influence as we move north thus lead to a higher frequency of basin water renewal. In fjords with similar characteristics, the residence time of the basin water in a fjord in western Norway will be approximately five times as long as in a fjord on the northern coast. The density of coastal water is usually at its highest in the winter and spring, when basin water renewal usually occurs.

If we consider a fjord with specific characteristics, whose basin water has been renewed by inflows of water with an oxygen content of 6.0 ml/l, this content will gradually fall in the course of time as a result of biological processes, reaching a minimum value that depends on the intensity of tidal vertical mixing. Figure 9.9 illustrates how increasing tidal mixing in fjord basins towards the north affects oxygenation. As we have seen, a typical fjord basin in the Skagerrak will reach an oxygen content of zero before the next inflow that renews the basin water. On the western coast, the minimum oxygen content of the basin water before the next inflow will be around 4 ml/l and on the northern coast, 5–5.5 ml/l.

Climatic changes

Roald Sætre, Jan Aure and Didrik S. Danielssen

10.1 Introduction

In addition to the short-term and seasonal variations in the physical conditions of the Norwegian Coastal Current described in previous chapters, there are also long-term interannual fluctuations. These are often referred to as ocean climate variability. Due to the energy exchange between atmosphere and ocean, the long-term climatic fluctuations in the atmosphere are reflected in the physical conditions of the ocean. The most conspicuous climate feature in the North Atlantic is the North Atlantic Oscillation (NAO), which is a large scale north-south oscillatory fluctuation in the atmospheric mass in the North Atlantic. The NAO index is expressed by the monthly pressure difference at sea level between the Azores High and the Iceland Low (e.g. Blindheim, 2004). Interannual fluctuations in the NAO index have a pronounced influence on both physical and biological conditions in the Nordic Seas. This is most clearly seen during the winter months, when atmospheric activity is most intense. Under high (positive) NAO conditions, the strong air pressure gradient between the Azores High and the Iceland Low drives southwesterly winds, producing mild wet conditions in northwestern Europe and dry conditions in the Mediterranean region. Under low (negative) NAO conditions the southwesterly winds are weaker, leading to dryer conditions in northwestern Europe and wetter conditions in the Mediterranean. Figure 10.1 shows the fluctuations in the NAO index for the winter months of December to March during 1930–2005.

▶ **Figure 10.1**
The NAO winter index (December–March) from the 1930s. The curve has been smoothed by calculating five-year running means.

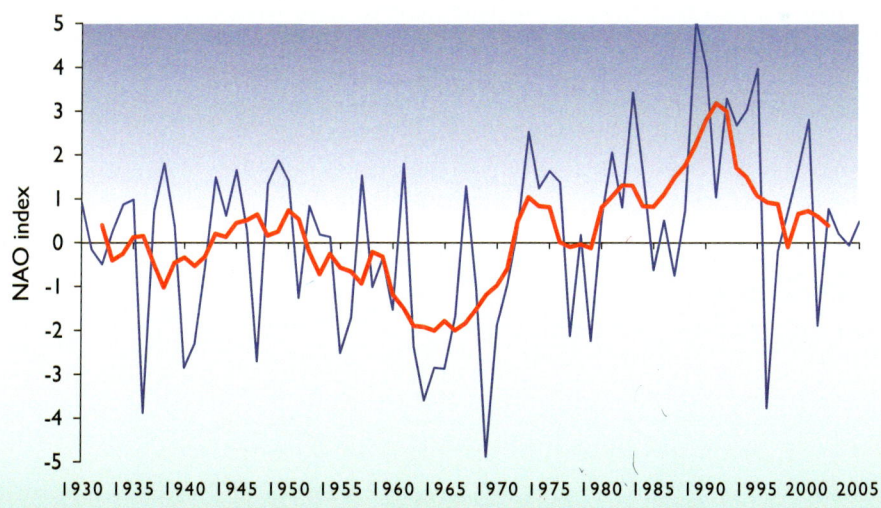

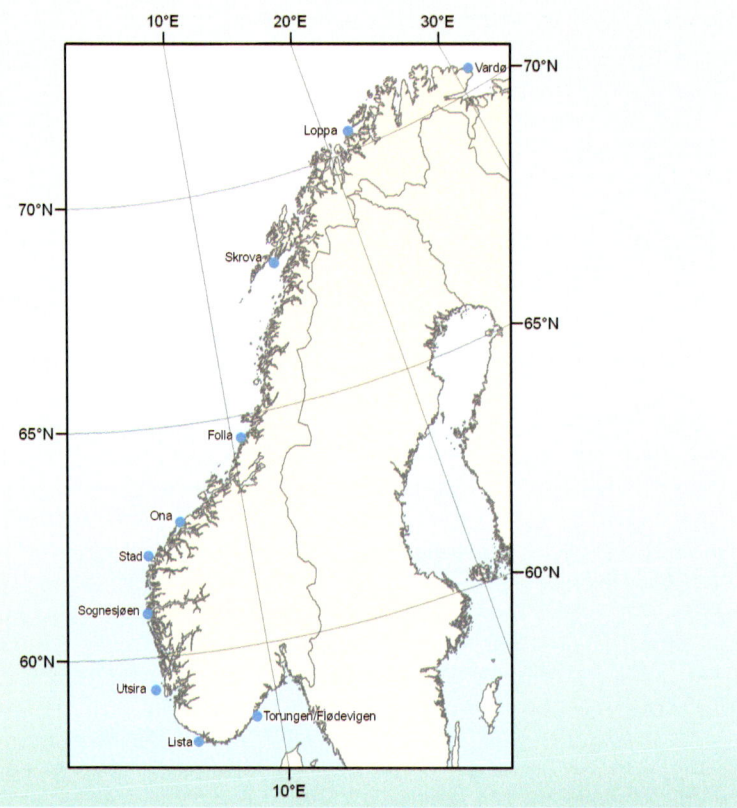

► Figure 10.2
The positions of the surface layer observation points (thermographic stations) and fixed hydrographic stations from which data have been used.

The longest time series of observations from Norwegian coastal waters comprises sea-surface temperature measurements from some lighthouses as a part of the meteorological observation system that started in about 1870. The Norwegian shelf provides spawning and hatching grounds for several commercially important fish species. Early in the previous century it was acknowledged that fluctuations in hydrographic conditions on the coast might influence the recruitment, growth and distribution of fish stocks. This was used as an argument for establishing the Norwegian Coastal Oceanographic Observing System (NCOOS) described in Chapter 2. NCOOS consists both of observations from the surface layer by ships of opportunity (the Thermographic Service) and of fixed hydrographic station observations made by local observers (Figs. 2.7 and 2.9). The system was established in the mid-30s and much of it is still operational. It represents some of the longest continuous oceanographic time series in the world. Figure 10.2 shows the positions of those observation points from which data in this chapter have been included.

Since 1919 temperature and salinity have been measured daily in the pipeline through which seawater is pumped from various depths at the IMR's Research Station in Flødevigen. Data from this observing system have been used in a large number of reports and publications for various purposes, including highlighting long-term variations (e.g. Ljøen and Sætre, 1978, Blindheim et al., 1981, Danielssen et al., 1996, Sætre et al., 2003).

In terms of climate the 1990s have been an exceptional decade. Both in the northern hemisphere and globally, this has been the warmest decade on record. The people of northwestern Europe have experienced this change in climate, particularly in terms of mild and rainy winters, but also with some record-high summer temperatures. Warming and extreme meteorological events of this sort may reflect global warming due to the release of greenhouse gases, in particular CO_2.

The aim of this chapter is to elucidate the long-term patterns of hydrographic variability on the Norwegian coast, and to relate these to global and North Atlantic regional climatic trends. The chapter refers primarily to the winter situation, as this season is believed to better reflect long-term climatic signals. As the climatic situation in the 1990s has differed from previous decades, we have particularly compared the 1990s with earlier periods.

10.2 Long-term fluctuations in the surface layer

At the meteorological stations in the lighthouses of Torungen and Ona (Figure 10.2) daily observations of the sea surface temperature have been made since 1867 and 1868, respectively. Figure 10.3 shows the smoothed fluctuations in mean winter (January–March) and summer (July–September) temperatures at these two stations since 1870. The two time series show similarities in the fluctuations, but their amplitude appears to be largest at Torungen. An identical variability pattern is found all along the coast (e.g. Blindheim et al., 1981), indicating that these fluctuations are due to large-scale climatic signals. The climate variations at Torungen are representative of the entire Skagerrak as well as of the central and northern North Sea (Ljøen and Sætre, 1978), while Ona represents the fluctuations on the coastline bordering the Norwegian Sea.

Figure 10.3 shows that during the winter there was no clear trend at Ona until the 1950s, while Torungen displayed a slowly rising trend during the same period. Between the 1950s and the 1980s there was a falling trend in the winter temperatures at both Torungen and Ona, followed by a distinct rising trend. The

▶ **Figure 10.3**
Mean surface temperatures for January–March and July–September at Torungen and Ona lighthouses since 1870. The curves have been smoothed by calculating five-year running means.

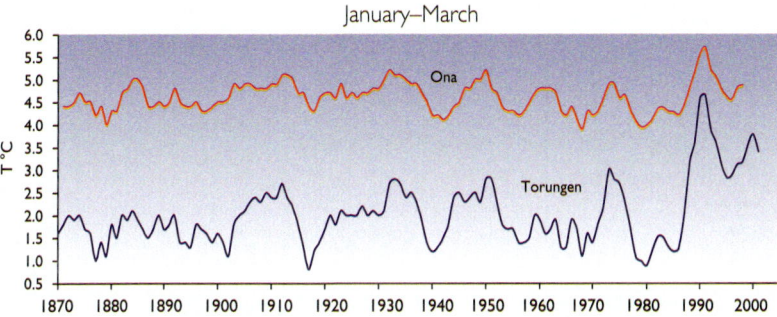

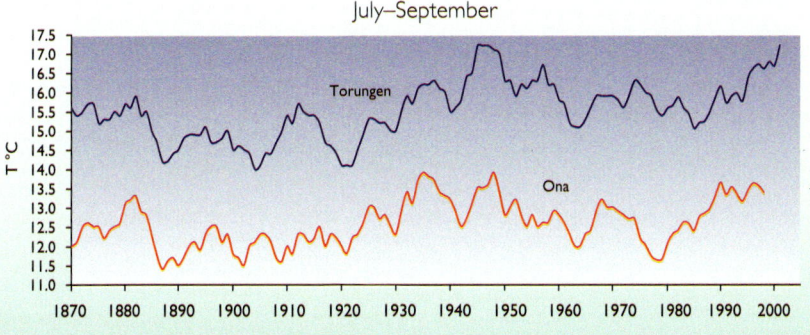

BOX 10.1 STORM SURGES ALONG THE NORWEGIAN COAST

▶ Satellite image of hurricane "Inga" in January 2005.

▶ Flooding of Bryggen in Bergen.

Storm surges or extreme high water levels occur as a combined effect of tidal and meteorological forcing. When strong winds from an atmospheric low are piling up water along the coast, and at the same time the atmospheric air pressure is low, a storm surge may occur. A difference in atmospheric pressure of 1 hectopascal (hPa) = 1 millibar (mb) equals about 1 cm in water level. A strong atmospheric low, for example of 960 hPa, would raise the water level by 50 cm compared to the normal atmospheric pressure (about 1010 hPa).

If this happens at full or new moon, and at spring tide, the water may reach extremely high levels and cause serious damage. Several such situations have been described in history, especially from the central and northern Norwegian coast. In January 1901 35 people drowned in a storm surge at Hærøy on the coast of Northern Norway.

On 12 January 2005 the hurricane "Inga" hit the coast of Western Norway, and in combination of strong southwesterly winds and very low atmospheric pressure, a storm surge developed. The combination of high water levels and large waves resulted in serious damage. Bridges and all seaward communication routes were closed down. Figure A shows a satellite image of hurricane "Inga" while Figure B shows the flooding of Bryggen in Bergen as one of its results. Figure C demonstrates the meteorological effects on the water level at Måløy (observed water level − calculated tide) during the storm surge and as can be seen, this approached 80 cm.

On 3–4 December 1999 an extreme atmospheric low passed the Skagerrak. The combined effect of low air pressure and wind resulted in a sudden drop in water level of about 150 cm in Oslo (Figure D). When the atmospheric low had passed, the water level rose by 220 cm in the course of five hours. Due to this extreme water level exchange, large water masses were forced into the Oslofjord, resulting in current speeds of about 5 m/s or 10 knots in the narrow parts of the fjord.

A future warmer climate will probably result in more frequent extreme meteorological situations, including higher waves and storm surges. The largest increases in the frequency and intensity of these are expected to be in the Barents Sea, while in the North Sea only minor changes are anticipated.

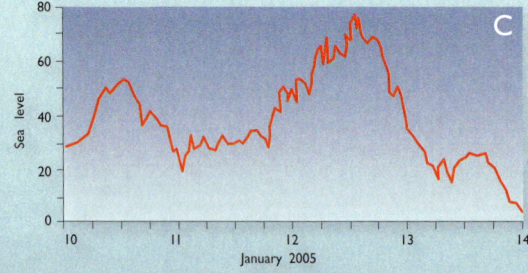

▶ The meteorological effect (wind and atmospheric pressure) on the water level at Måløy during hurricane "Inga".

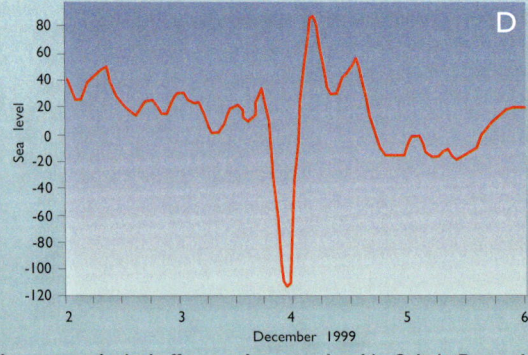

▶ The meteorological effect on the water level in Oslo in December 1999.

winter temperatures show pronounced variations of 1–2° C between warm and cold periods at intervals of 10–20 years. A characteristic feature is the warm winters around 1910, 1930, 1950, 1970 and 1990. Figure 10.3 shows that the winter of 1990 was the warmest winter of the past 130 years at both Torungen and Ona.

During the summer months (July–September) there has been a pronounced rise in temperature from the last part of the 19th century up to around 1950, followed by a falling trend until about 1980, since when summer temperatures have risen considerably.

Figure 10.4 shows mean and smoothed temperatures for the first quarter of the year (January–March) at a depth of 10 m at the fixed hydrographic stations Utsira and Skrova from 1940 to 2005 (Figure 10.2). Between 1940 and 2000, four relatively warm winter periods could be identified, peaking around 1950, 1960, 1975 and in 1990–1992. From around 1950 until 1985 there was a negative temperature trend along the whole coast, followed by a generally positive trend, as can also be seen in Figure 10.3.

The last period of warm winters started in 1988 and peaked at the beginning of the 1990s. During this period the highest winter temperatures since 1867 were observed along the southern and central parts of the Norwegian coast. Further north, the rise in temperature in the 1990s was significantly less, as can be seen in the observations from Skrova (Figure 10.4). Comparing the decadel mean temperature for the 1990s with the mean for 1940–1989, the deviation falls from around 1.5 °C at Flødevigen to 0.1 °C at Vardø (Figure 10.5).The effect of the warm 1990s was thus gradually lessened from south to north.

In the surface layer there is also a significant negative trend in winter salinity along the whole coast until the end of the 1980s, with the strongest fall in salinity at the end of this period, as can be seen at Utsira in Figure 10.6. The trend was

▶ **Figure 10.4**
Mean temperatures for January–March at a depth of 10 m at the fixed hydrographic stations of Utsira and Skrova. The curves have been smoothed by calculating five-year running means.

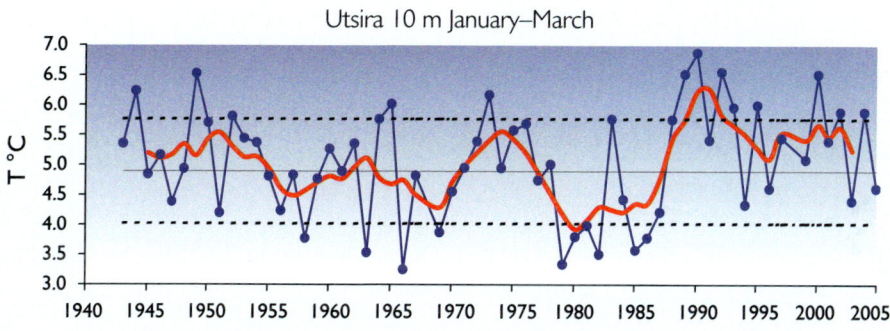

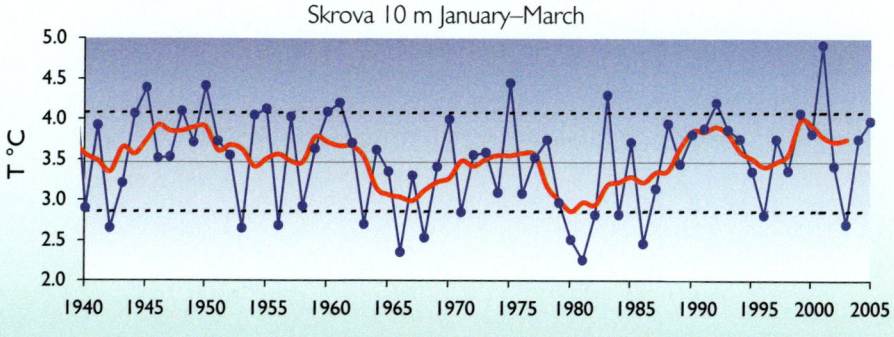

BOX 10.2 EFFECTS OF EXTREME WINTER

▶ View from the open Skagerrak towards the Torungen lighthouse during the winter of 1942

Photo: Birger Dannevig

▶ Vessels with ice problems on the Skagerrak coast during the winter of 1970

Photo: Birger Dannevig

As Figure 10.3 shows, there have been pronounced differences between cold and warm winters at intervals of about 10 to 20 years. Examples of extremely cold winters are 1940–42, 1947, 1963, 1966 and 1970. These years are characterised by air temperatures well below the normal for relatively long periods. During these extreme winters sub-zero temperatures in the surface layer could be found throughout Skagerrak, as well as in the coastal water north to Utsira (Figure 10.2). On the Norwegian Skagerrak coast, both locally formed and drifting ice from the Kattegat could be a serious problem for shipping (Figures A and B).

It appears that such extremely cold winters result in cooling and water mass exchange in the deeper layers of the coastal region as well. Eggvin (1943) described the great exchange of water masses on the Norwegian coast during the cold winter of 1940. Two mechanisms appear to be involved: 1. Relatively high-salinity water on the outer parts of the continental shelf is cooled and sinks to replace the deeper layers in the near-coast areas. 2. Relatively high-salinity water in the deeper layers of the near-coast areas and in the fjords is brought to the surface by wind-induced upwelling and turbulent mixing, where it is cooled and subsequently sinks.

The water masses of the northern North Sea shelf are vertically well mixed and homogeneous during the winter, and the cooling process reaches the bottom. During cold winters the bottom temperature will thus be very low, and this situation will persist until the following winter. During these winters, due to the exceptional cooling, the northern North Sea shelf water may become much heavier than the water of the deeper parts in the Norwegian Trench. This cold shelf water will then cascade down the slope of the trench to renew the basin water of the Skagerrak Deep, causing a deep-water temperature drop of about 2 °C (Figure C). Transport of heat from above raises the temperature again, but it takes several years to reach its former level (Ljøen, 1970).

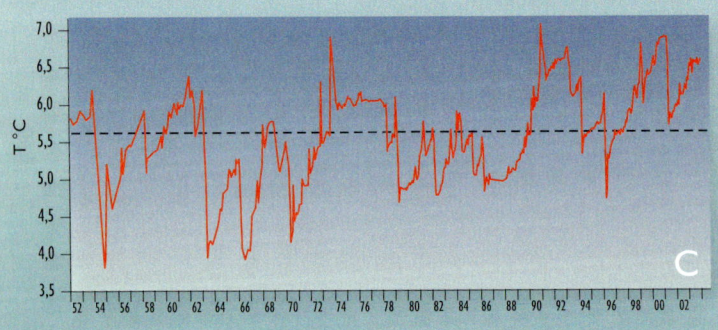

▶ Bottom water temperatures (600 m) in the Skagerrak for 1952–2004.

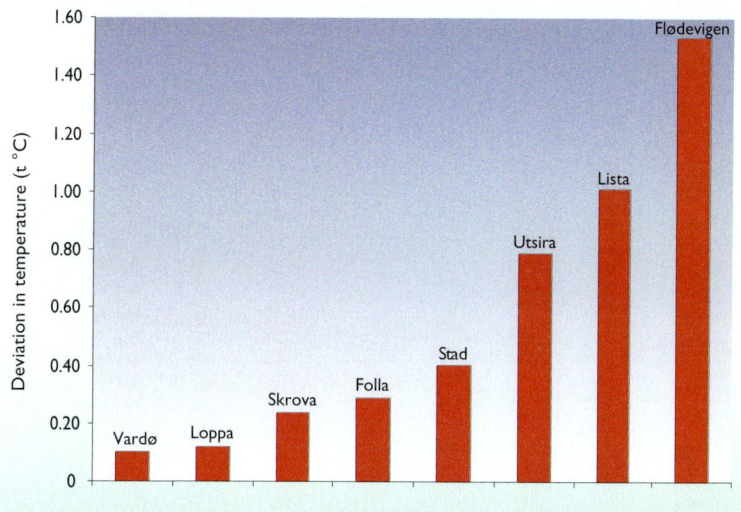

▶ Figure 10.5
Deviations between the decadal mean winter temperature for the 1990s and the mean temperature for 1940–1989 in the surface layer at selected coastal stations.

▶ Figure 10.6.
Mean salinities during January–March at a depth of 10 m at the fixed hydrographic stations of Utsira and Skrova. The curves have been smoothed by calculating five-year running means.

most marked at the coastal stations in open waters influenced by the Atlantic inflow, while it was considerably less at stations more affected by the fjords, such as Skrova. The general fall in salinity up until the 1980s was most marked along the southern and central coast, and gradually decreased northwards along the coast. From the end of the 1980s on, the salinity of the surface layer rose once again, as seen at Utsira in Figure 10.6, while the negative trend continued at Skrova. Like the temperature records, the salinity increase in the 1990s was markedly higher along the southern coast. Only from Utsira and southwards were the salinities in the 1990s above the 1940–1989 mean value (Figure 10.7).

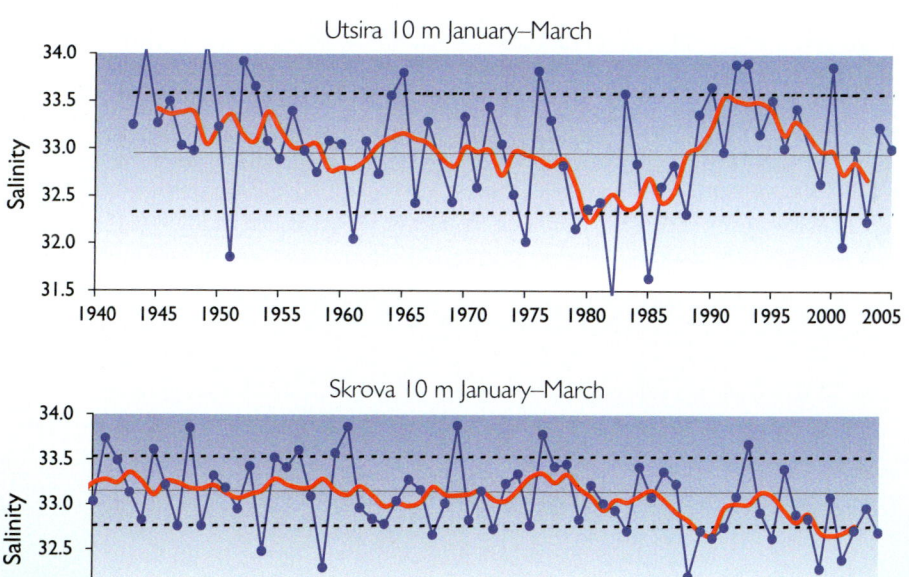

Climatic changes 131

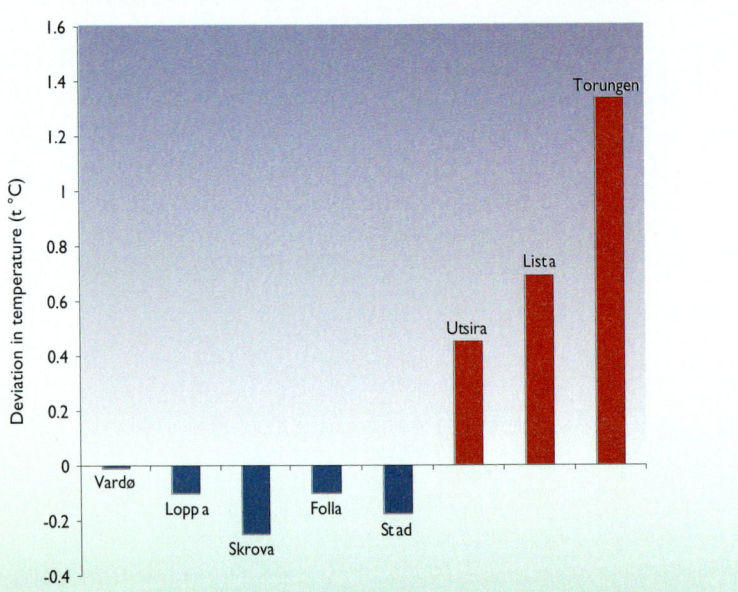

▶ Figure 10.7
Deviations between the decadal mean winter salinity for the 1990s and the mean salinity for 1940–1989 in the surface layer at selected coastal stations.

▶ Figure 10.8
Mean temperatures 1960–2005 at a depth of 1 m at Flødevigen for August and March.

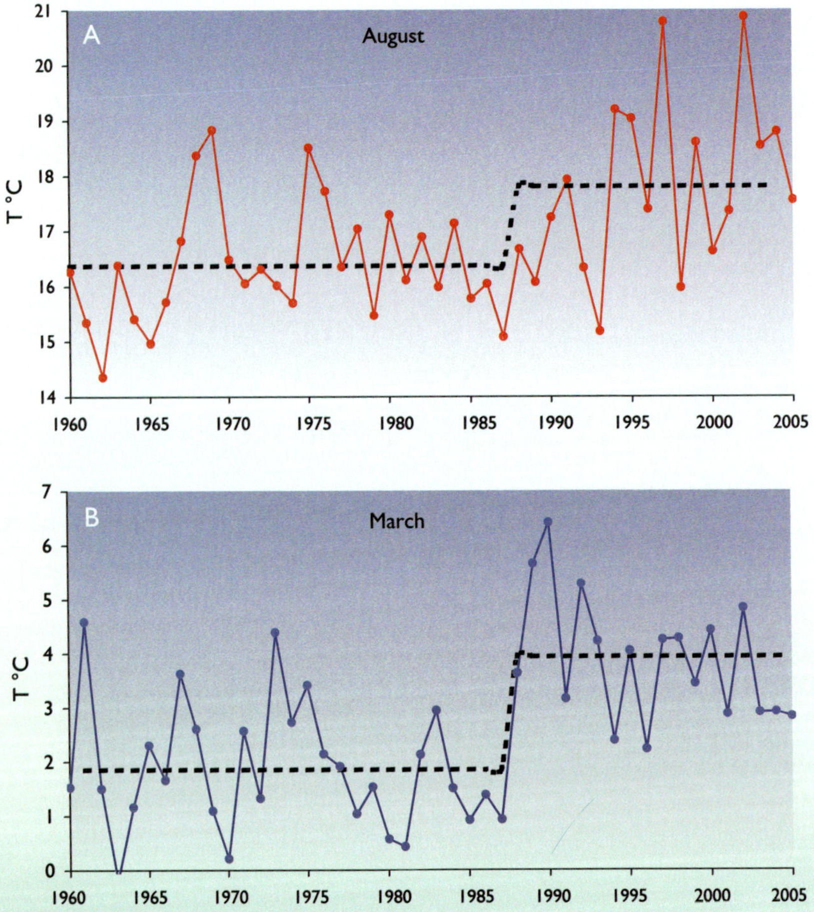

On the southern coast there appears to have been a shift in the temperature regime around 1988, with a sudden jump in mean summer and winter temperatures, after which temperatures continued to vary, although now at a higher level. This feature is especially pronounced in the Skagerrak (Figure 10.8). Since 1988 there have been a number of warm winters along the Skagerrak coast, with extremely high winter temperatures in 1989 and 1990. The period since 1988 has been the warmest for the past 130 years (Figure 10.3). After approximately normal winters in 1994 and 1996, the warm winters continued, with temperatures well above the normal. There have also been several warm summers since 1990, with the summers of 1997 and 2002 the warmest since observations began in 1924. If we compare the seasonal mean temperatures for 1940–1987 with 1988–2000 we can see that there has been a sudden increase of about 2 °C in winter and 1.5 °C during the summer (Figure 10.8).

10.3 Long-term fluctuations in the deeper layer

In the deeper layer (150–200 m) of the coastal water too, there has been a negative temperature trend from about 1945 until the beginning of the 1980s, as exemplified by Figure 10.9. After the cold period around 1980, temperatures increased in 1990–1992 to reach their highest level since observations began in 1936. After a marked temperature drop in 1993–1994 there has been a further gradual rise in temperature.

In the deeper layer there has also been a general decrease in salinity from around 1960 until the early 1980s (Figure 10.10). At the stations off the southern coast the fall was highest from 1970 to the early 1980s, while further north at Skrova there was a gradual decrease from the mid-1960s until the beginning of the 1980s. The increase in salinity from the mid-1980s and in the 1990s was highest at the stations that are most heavily influenced by Atlantic Water, such as Utsira. The relatively cold and low-salinity period at the end of the 1970s

▶ **Figure 10.9**
Mean temperatures during January–March at a depth of 150 m at the fixed hydrographic stations of Utsira and Skrova. The curves have been smoothed by calculating five-year running means.

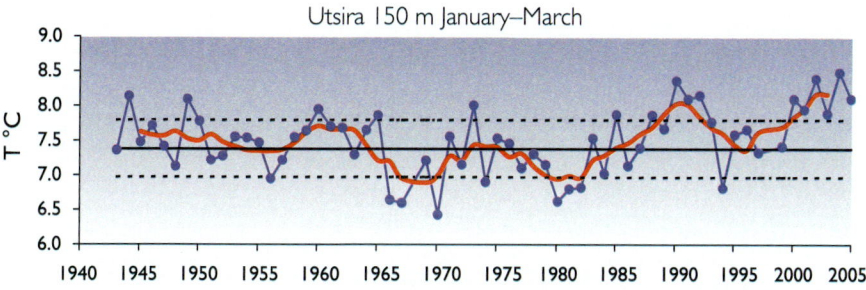

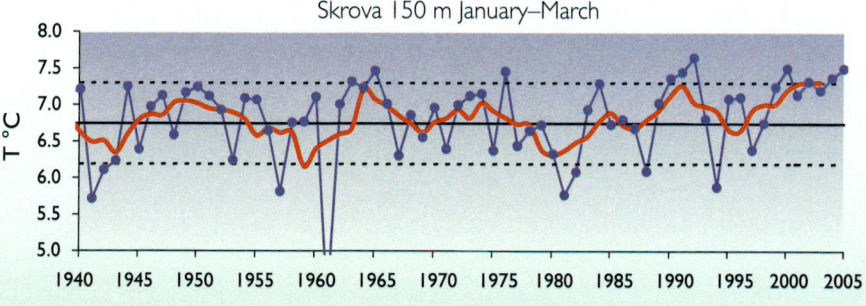

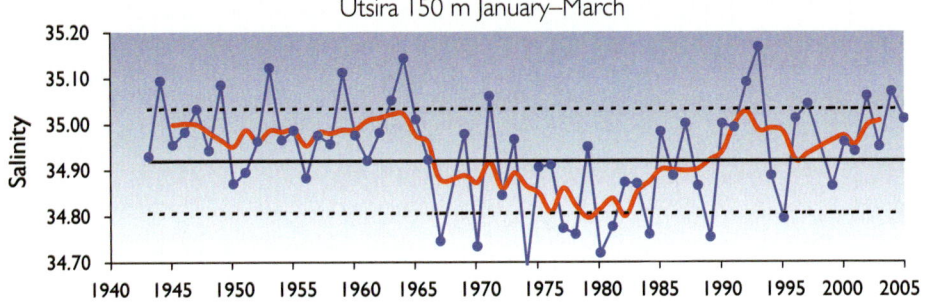

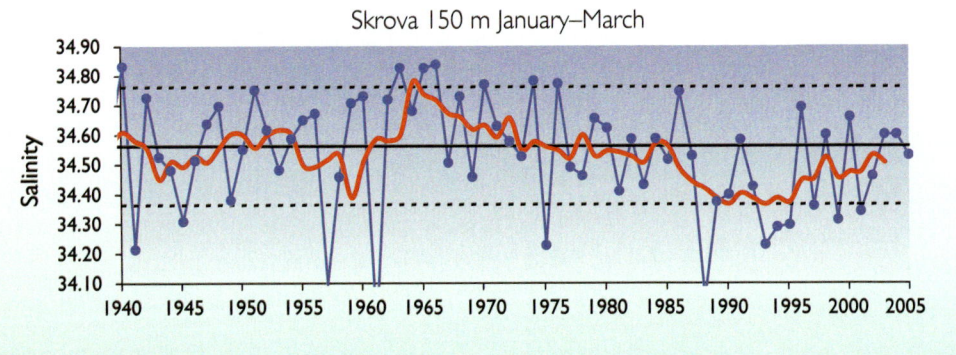

▶ Figure 10.10
Mean salinities during January–March at a depth of 150 m at the fixed hydrographic stations of Utsira and Skrova. The curves have been smoothed by calculating five-year running means.

and early 1980s appears to be associated with a well-known outbreak of polar water that influenced the whole North Atlantic including the North Sea and the Norwegian Coastal Current and is referred to as the "Great Salinity Anomaly" (Dickson et al., 1988).

10.4 What cause the climate fluctuations?

The variable inflow of Atlantic Water to the Norwegian and North Seas influences the hydrographic conditions on the coast of Norway and the Skagerrak. The modelled winter inflow to the North Sea during January–March from 1955 to 2000 is shown in Figure 10.11. As the large-scale wind pattern is the main driving force, the fluctuations in this inflow are likely to reflect the variability in the inflow to the Norwegian Sea. As can be seen, there was a significant increase in the inflow in the late 80s and 90s. Between 1988 and 2000 there were six winters with extremely high rates of inflow. The mean inflow in the 1990s was 1.7 SV, which is more than 20 % above the long-term mean for 1955–1989. It has been shown that this inflow was highly correlated with the North Atlantic Oscillation (NAO) winter index (Figure 10.1). A high NAO index is associated with mild winters and an increase in westerly winds and winter precipitation over Scandinavia.

As mentioned in Chapter 5, the wedge-shaped Norwegian Coastal Current is deep and narrow during the winter and wide and shallow in summer. The driving mechanism for this seasonal lateral oscillation of the current is probably an effect of the monsoon-like wind pattern along the coast of Norway (Sætre et al., 1988). High NAO levels mean stronger southwesterly winds and this will deepen the Norwegian Coastal Current off most of the coast and could result in colder and less saline upper coastal water. Higher wind speeds will also have an

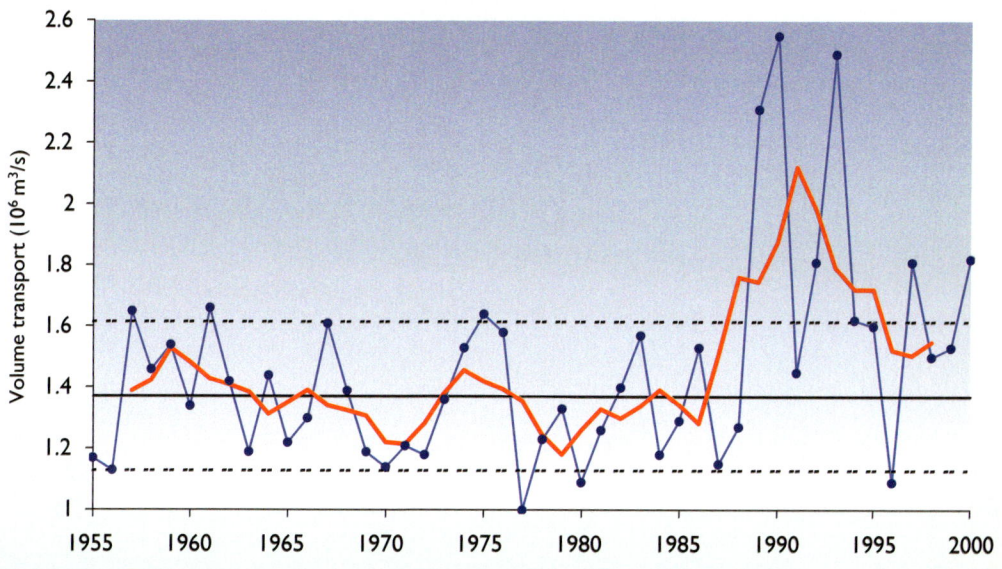

▶ Figure 10.11
The North Sea winter inflow (January–March) between Utsira and the Orkney Islands, simulated by a numerical model.

effect on the vertical mixing of water masses and on local cooling. A high NAO winter index is not therefore necessarily synonymous with higher temperature and salinity in coastal waters.

The higher winter temperatures and salinities in the Norwegian coastal water in the 1990s are clearly related to the significant increase in the inflow of relatively warm highly saline Atlantic Water during the same period, particularly in the deeper layers, which are more directly influenced by the Atlantic Water. The pattern of variability in the Atlantic inflow to the Norwegian Sea is probably quite similar to that of the North Sea (Figure 10.11). In the deeper layers of the coastal water (150 m) a clear Atlantic influence can be observed at all coastal stations. In the upper coastal water (10 m), however, the Atlantic influence is most pronounced in southern Norway, with a clear decrease as we move north.

The long-term falling trend in salinity in the coastal water off southern Norway is probably primarily caused by increased precipitation and thereby the substantial increase in freshwater runoff. Here there are differences between stations strongly influenced by the fresh water outflow from fjords and stations at which oceanic influences are more dominant. Førland et al. (2000) demonstrated that winter precipitation in western Norway was more than 25 % higher in 1980–1999 than during the normal period 1961–1990. This pattern of precipitation probably reflects that of northwestern Europe, so that during the 1990s, the Norwegian Coastal Current has been supplied with more fresh water from both the Baltic and the North Sea. One of the effects of the regulation of the fresh water runoff from hydroelectricity schemes is increased winter discharge, and this may also be an important explanatory factor. Asvall (1976) showed that in southeastern Norway the mean natural winter fresh water discharge during the period 1969–1973 rose by as much as 170 %, due to flow regulation.

On the southern and central coast the winters of the 1990s were characterised by the highest decadal mean temperature in the surface and deeper layers for the whole period of observations. In the upper layer this tendency was most pronounced in the southern parts of this sector. In the upper layer of the northern coast, however, other decades, such as the 1950s and the 1970s, show higher decadal mean temperatures.

On the northern Norwegian coast bordering the Barents Sea the patterns seem to follow that observed in the open Barents Sea (Ingvaldsen *et al.*, 2001),

BOX 10.3 CLIMATE FLUCTUATIONS AND FISH STOCKS

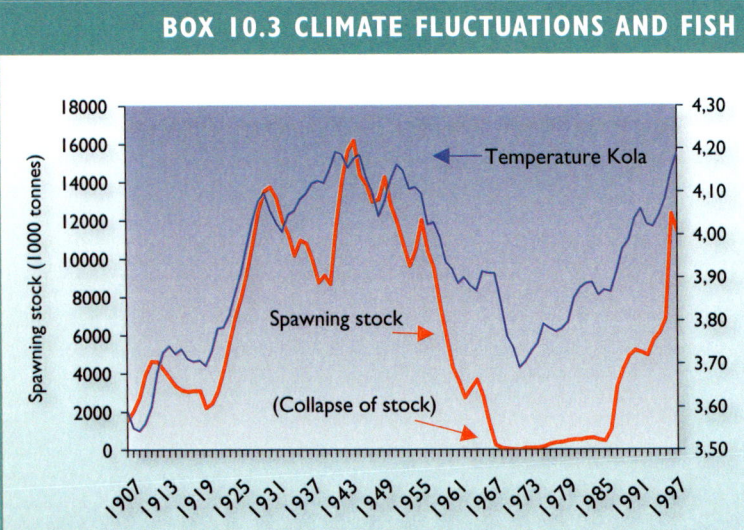

▶ The long-term variations in the annual mean temperature in the Kola section in the Barents Sea together with the calculated spawning stock size for the herring for the previous century (Toresen and Østvedt, 2000).

In ecosystems at higher latitudes, sea temperature is a key parameter for growth and recruitment to fish stocks. However, the apparent temperature effects have the potential to camouflage mechanistic relationships with other climate parameters that correlate to some extent with the sea temperature. These may include various effects of wind and variable current and light conditions, as well as precipitation and evaporation.

The size of the fish stocks on the Norwegian coast has varied considerably in the course of time. History tells of periods when cod and herring were present in great amounts, followed by times of absence for several years or, in the case of the herring, even decades. The stock history of the Norwegian spring-spawning herring demonstrates this. Explaining these natural fluctuations in the fish stocks was a key challenge for the pioneers in marine research a century ago.

The figure shows the long-term variations in the annual mean temperature in the Kola section in the Barents Sea together with the calculated spawning herring stock size for the previous century. The Kola section is used because it is the longest consistent oceanographic time series that reflects climate variability in our region. As can be seen, there was a relatively cold climate at the beginning of the last century. Temperatures rose during the 1920s and 1930s, with a subsequent decrease to a minimum around 1970. Later, the temperature again increased towards the warm period of recent years. The fluctuations in the spawning stock of herring vary approximately in the same way as the temperature, demonstrating a relationship between ocean temperature and the quantity of herring. The actual mechanism for such relationship is not quite clear, as temperature could be a proxy for other climatic parameters. However, the most likely explanation is that high temperatures indicate favourable conditions for recruitment to the fish stocks.

In the 1960s new technology made the fishery more efficient, resulting in increased fishing pressure, particularly on the juvenile herring. The collapse of the herring stock in the late 1960s was probably a combined effect of excessive fishing pressure and unfavourable conditions for new recruitment to the stock.

Periodic fluctuations in the global climate are a natural process. The possible effect of human influence on the climate will add to the natural variations and this is expected to result in a significant temperature increase in the near future. Such a temperature increase could have a rather dramatic effect on fish stocks, especially on their distribution. In general, a northeasterly displacement of the most important fish stocks is expected and new species originating from further south will be more common off the coast of Norway.

where the 1930s and 1950s in particular were warmer than the 1990s. Although the NAO has a significant effect on the climatic variability of the area, local winds and differences in atmospheric pressure between the Norwegian and the Barents Seas seem to be major factors that determine both the degree of Atlantic inflow and circulation and water mass distribution within the Barents Sea (Ingvaldsen et al., 2001).

Since 1990, there has been a falling trend in the NAO index, and since around 2000 it has been back to about its normal level. In spite of this, however, winter surface temperatures since 2000 have been above the normal along the whole coast and there has been a strong increase since 1995, in both the temperature and salinity of the deeper layers (e.g. Figures 10.9 and 10.10). Long-term observations of the Norwegian Atlantic Current since 1995 show a falling trend in the current speed and thereby in the water volume transported. The temperature in the core of the inflowing water, however, has risen by 1 °C (Orvik and Skagseth, 2005). The Atlantic inflow salinities along the routes into the Nordic Seas were record-high in 2004. At Ocean Weather Station Mike (66°N, 2°E) in the Norwegian Sea, the salinity of the Atlantic Water in 2004 was the highest observed since observations started in 1948 (Gammelsrød and Hjøllo, 2005). Hátun et al., (2005) conclude that the interannual to decadal salinity fluctuations of the Atlantic inflow to the Nordic Seas are primarily controlled by the intensity of the cyclonic Subpolar Gyre (SPG) in the North Atlantic south of the Scotland–Greenland Ridge, which is directly related to the NAO index. During weak circulation intensity, i.e. low NAO indices, the SPG is displaced towards the west opening up for greater inflows of saltier and warmer water from the eastern North Atlantic.

The NAO index is a simple expression of the atmospheric pressure difference between the Icelandic Low and the Azores High. It is only a convenient large-scale index for atmospheric conditions in the North Atlantic and it does not include important details, such as an east-west displacement of two pressure extremes. It appears that during the winters of the late 1990s the Icelandic Low was displaced slightly towards the east (Dickson and Meincke, 2003).

The observations of long-term temperature and salinity variability in Norwegian Coastal Waters suggest that the fluctuations in the upper coastal water are caused both by advected properties such as the Atlantic inflow, and by local processes, including heating and cooling, fresh water runoff and local winds. The variations in the deeper part of the coastal water are mainly caused by the properties and amounts of Atlantic water transported into the area.

10.5 What of the future?

In recent years, a number of time series analyses have told us that we have entered a period of global climate change and that we are approaching warmer conditions. A growing number of scientists are now convinced that the heating trend reflects human influence on the climate through the release of greenhouse gases such as CO_2.

Even under conditions of global warming there have been arguments as to what will happen in Northern Europe, particularly with regard to the possible fate of the Gulf Stream or the Atlantic inflow to the Nordic Seas. One of the main driving forces of this inflow is the formation of deep and bottom water in the Greenland Sea, where the high-salinity Atlantic Water is cooled, sinks and flows southwards to leave the Nordic Seas across the sill between Scotland and Greenland. Global warming could result in more melting of ice in the polar region and thereby an increased supply of fresh water to the Greenland Sea. The process of deep and bottom water formation may then be retarded or even blocked, which in turn could significantly reduce the Atlantic inflow to the Nordic Seas and thereby heat transport to northern Europe. A consequence of such a situation could be a colder climate in our region, in spite of a general global warming.

Other studies indicate that vertical mixing in the ocean, such as that generated by the tides, may supply energy for the continued formation of deep and bottom water in the Nordic Seas. The general wind pattern over the North Atlantic Ocean is another important driving force for the Atlantic inflow to the Nordic Seas. One consequence of global warming will probably be greater wind activity over the North Atlantic Ocean, which will encourage increased inflows of Atlantic water. The climatic development of the two most important driving forces of the Atlantic inflow to the Nordic Seas is thus likely to have the opposite effect. The two processes are not independent and other processes may also play a role. The climate change expected to take place over the Northeast Atlantic will therefore be the result of complex interactions among all these factors. The probability of a weakening of the Gulf Stream and thus of a reduced Atlantic inflow to the Nordic Sea is therefore still a matter of discussion among scientists.

A Norwegian research programme, RegClim, has been in operation for several years. The general aim of this programme is to produce scenarios for regional climate change suitable for impact assessment in northern Europe, bordering sea areas and major parts of the Arctic, given a global climate change. The programme will also quantify uncertainties due to the choice of methods, global scenarios and poorly understood processes that influence the climate of our region, in particular those that leave the Nordic Seas warm and ice-free. RegClim has produced a weather prediction for possible climate change in Norway for the next 100 years, which can be summarised as follows:

- the climate will be warmer, wetter and somewhat dryer, with slightly more winds and a higher probability of extreme weather events.
- The annual mean temperature will increase by 2.5–3.5 °C. Winter temperatures will rise, particularly in northern Norway.
- Summer temperatures will rise most in southern Norway.
- Annual precipitation will rise by 5–20 %, mainly during the autumn, with the highest increase along the southwestern coast and the northern coast.
- Smaller increase in wind speeds with increased frequency of extreme wind events.
- Fewer modifications of the circulation of the North Atlantic than predicted by other models. Formation of deep and bottom water will be reduced by 5–15 % when CO_2 content is doubled.

Operational oceanography – challenges and potential

Johnny A. Johannessen, Bruce Hackett, Einar Svendsen, Henrik Søiland, Lars P. Røed, Nina Winther, Jon Albretsen, Didrik Danielssen, Lasse Pettersson, Morten Skogen and Laurent Bertino

11.1 Introduction

Operational oceanography is for the ocean what meteorology is for the atmosphere; the activity of routinely making, disseminating and interpreting measurements of the seas so as to;

- Provide continuous forecasts of the future condition of the sea, as far ahead as possible (*Forecasts*).
- Provide the most usefully accurate description of the present state of the sea, including its living resources (*Nowcast*).
- Assemble climatic long-term data sets which will provide data for descriptions of past states and time series showing trends and changes (*Hindcast*)."

Operational oceanography is clearly multi-disciplinary in nature, covering as it does physical and biochemical oceanography, sensor development and observation methods, data processing and dissemination, numerical modelling and data assimilation.

It has long been known that ocean physics is a key driver of marine ecosystem variability. To the best of our knowledge, the first effort to use such information operationally was made by Jens Eggvin at the Institute of Marine Research (IMR) in Bergen. Eggvin who established an ocean observing system in 1935 (still ongoing), provided regular reports on the status and events of the physics of the ocean (particularly temperature and salinity), and explored the feasibility of forecasting the impact of ocean physics on fisheries. This an activity is also known as fisheries oceanography (see Chapter 2).

In the wake of the harmful *Chrysocromulina polylepis* bloom in May–June 1988 (Dundas *et al.*, 1988), which created serious problems for the aquaculture industry, the Ministry of Environment established a pilot programme (Havovervåking og Varsling – HOV) on ocean monitoring and forecasting (e.g. Johannessen and Pettersson, 1988). A significant cooperative effort between several Norwegian research and management institutions was made during 1990–1994 to build a national centre in Bergen for such activity. The work was hindered by organisational problems as some of the participating institutions felt that the role and the responsibility of the different partners was unclear. A passive Ministry of Environment took no correctional action and the program was unfortunately terminated at the end of 1994.

Although Helland-Hansen and Nansen (1909) touched upon hydrographic variability at different time scales, it was the emergence of satellite imagery in the 1980s which for the first time demonstrated that meanders and eddies are major features of the inflowing Atlantic Water to the Nordic Seas as well as of the Norwegian Coastal Current (e.g. Johannessen et al., 1986; Johannessen et al., 1989). At that time, two-dimensional ocean models were run operationally at the Norwegian Meteorological Institute to provide current and sea level information. However, to simulate frontal processes and mesoscale variability as depicted by the satellite images, fully prognostic three-dimensional models of the physics were clearly needed.

Such models were imported to Norway in the late 1980s, when the offshore industry became active on the continental shelf, and funded the MetOcean MOdelling Project (MOMOP) (Røed et al., 1995, Hackett et al., 1995). Capitalizing on this, the three-dimensional Princeton Ocean Model (POM) (Blumberg and Mellor, 1987) was selected as the first operational ocean model at the Meteorological Institute in the mid-1990s (Engedahl, 1995; Martinsen et al., 1997). This was the forerunner of today's forecast model (MIPOM) (Engedahl et al., 2001), producing daily updated forecasts of currents, temperature and salinity. In the 1990s, IMR extended models of the physics with chemistry and biology to the Norwegian Ecological Model (NORWECOM) system, yielding estimates of nutrients and primary production in the North Sea and Skagerrak (e.g. Skogen et al., 1995). Similar tools were also developed for the Barents Sea (e.g. Slagstad and Støle-Hansen, 1990). In 2001 MIPOM was coupled with NORWECOM to make the first operational system for prediction of the chemistry and biology of a major ocean area including the North Sea and Norwegian coastal waters.

At international level, the recognition of new ocean modelling and observational tools stimulated the establishment of the Global Ocean Observing System (GOOS) in the 1990s. In Norway we became involved in the European component of GOOS (EuroGOOS), in particularly the North West Shelf Operational Oceanographic System (NOOS) and Arctic GOOS sub-groups. International agreements, such as the Oslo–Paris (OSPAR) Convention targeting the North-East Atlantic and North Sea and the more recent EU Water Framework Directive, (see Chapter 1), place obligations on nations to monitor, protect and report on the status of the marine environment. At the turn of the century, the Global Monitoring System for Environment and Security (GMES) emerged as a joint EU and European Space Agency (ESA) initiative to provide a common infrastructure for monitoring needs. With a target date of 2008 for operational implementation, it has been proposed that GMES Marine Services should support and strengthen European capacities for:

- verifying and enforcing international treaties and informing the assessment of policies
- enabling sustainable exploitation and management of ocean resources (offshore oil and gas industry, fisheries, aquaculture industry)
- improving the safety and efficiency of maritime transport, shipping and naval operations, supporting national security and reducing public health risks
- anticipating and mitigating the effects of environmental hazards and pollution (oil spills, harmful algal blooms)
- advancing marine research for better understanding of the ocean ecosystems, their variability and contribution to climate change issues
- contributing to seasonal climate prediction and mitigating its effects on coastal populations
- supporting specific services for coastal management and planning.

Several EU research and development projects, such as Towards an Operational Prediction System for the Atlantic European Coastal Zones (TOPAZ) have been tailored to developing operational systems describing the physics of the ocean in response to many of the above-mentioned initiatives and programmes. These have usually focused on global to regional perspectives. TOPAZ covers the North Atlantic, Nordic Seas and Arctic Ocean, and assimilates satellite-derived sea surface temperature, sea level anomalies and sea ice concentration and extent. Daily best estimates and weekly forecasts, including three-dimensional temperature, salinity and current fields (Bertino and Evensen, 2003) are routinely delivered by the Mohn-Sverdrup Center/NERSC. The Meteorological Institute runs its own system with daily updated seven-day forecasts for the Northeast Atlantic and different parts of the Norwegian coastal areas. A more regional- to coastal-scale European project, is about to start. The Norwegian contribution will focus on the North Sea and the Norwegian Coastal Current. It extends parts of the development and operation of the data- and model-based system implemented under the Monitoring the Coastal Zone Environment (MONCOZE) project (Johannessen et al., 2006) that was supported by the Research Council of Norway between 2001 and 2005 (see Chapter 2).

Norway has been a forerunner in including chemical and biological oceanography in its operational information services, and sophisticated ecological models and improved observation technologies are now gradually emerging. However, resources are lacking for implementing an integrated and coordinated operational monitoring and forecasting system capable of producing useful information on the dynamics of our marine and coastal ecosystems (e.g. for fisheries predictions, assessment and management advice). A national plan (Svendsen et al., 2002) was presented for the development of such a model-based tool to integrate existing and new multi-disciplinary knowledge and data from physics to mammals into an innovative system for assessing the historical, current and future states of the marine ecosystems within our exclusive economic zone. Unfortunately the plan was not approved in its full breadth for financial support. However, the ten-year programme (Havet og Kysten), initiated by the Research Council of Norway in 2005, is a step in the right direction.

11.2 Challenges

The Norwegian coastal and offshore regions enjoy some of the richest fisheries in the world. In addition, the many fjords and inlets in the area provide the basis of one of the world's largest aquaculture industries. A significant percentage of the world's oil and gas is also produced in the area (see Chapter 1). The Barents Sea and the Siberian Shelf are estimated to have about 20 % of estimated global remaining oil and gas reserves.

Reliable forecasts of sudden events, such as extreme currents, harmful algae blooms, or drift of oil spills and descriptions of trends and variability of the state of the ocean, are of crucial importance for these industries. Ocean predictions are also of great importance for ongoing monitoring of the marine environment, and their value to the scientific community should not be underestimated. Moreover, forecasts of ocean state are an essential decision-support tool for search and rescue operations, the management authorities, the shipping industry, and the safety and health of the general public. This field also encompasses predicting the effects of management actions (what-if scenarios) such as changes in nutrient loads.

Most of the extreme and hazardous currents and hydrographic events in the ocean are associated with the generation and propagation of mesoscale features such as eddies, meanders and jet filaments. This is akin to the atmosphere and constitutes the "ocean weather." The Norwegian Coastal Current is a prime

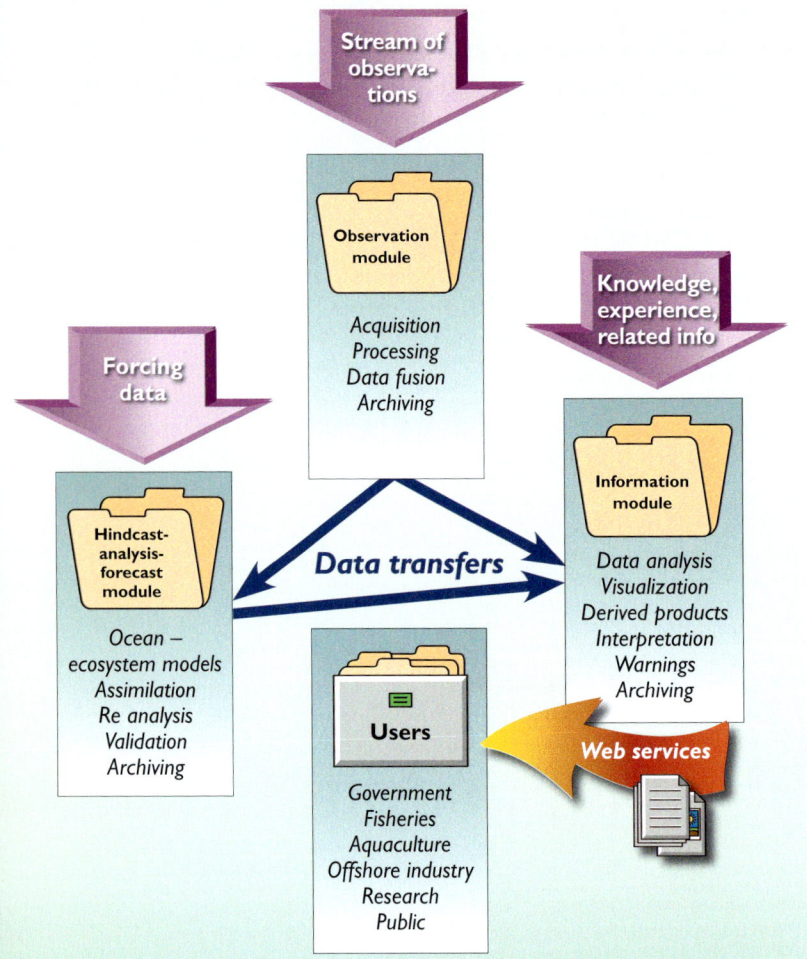

► Figure 11.1
Schematic of ocean monitoring system such as developed for MONCOZE. The Observation module builds largely on existing data streams from operational satellite and *in-situ* observing systems, but must accommodate new observations as they become available. The Hindcast-analysis-forecast module relies on numerical models for ocean circulation and ecosystem that produce operational forecasts as well as state descriptions and scenarios. Observational and model data are analyzed and interpreted on the basis of scientific knowledge in the Information module; dissemination of information and data are facilitated by web services. The system encompasses both real-time and delayed mode information.

Graphics: Harald E. Tørresen

example, in which such mesoscale features are abundant (Røed and Fossum, 2004). The dynamic evolution of the ocean weather, however, is much slower than that of the atmosphere. At the same time the spatial length-scale is much smaller. Furthermore, *in-situ* ocean observations available in near-real-time are scarce, which suggests that making accurate predictions of the ocean weather would present a number of scientific challenges. Among these are: what types of observations are essential, what is the best use of the available observations, what is the current state of the ocean, what are the uncertainties in the forcing fields, how large are the observation and model uncertainties, and how far into the future is it actually possible to predict ocean weather, both theoretically and in practice?

Another prominent challenge is that of understanding and quantifying the impact of ocean weather on biology. The physical parameters of the ocean frame crucial biological processes such as photosynthesis, migration, recruitment and growth of biological organisms. By improving the prediction of ocean weather, the accuracy of simulations of important biological processes is expected to be improved. An adequate monitoring system must rely on near-real-time availability of routine *in-situ* and satellite-based observations of the current field, thermohaline structure and lower-level trophic conditions, together with access to atmospheric forcing fields and robust, operational predictive ocean models. Open boundary conditions in these models need to be realistically treated with respect to water mass exchange by downscaling from wide area coverage, coarser resolution models. Estimates of river discharges and nutrient loads are also required. Accurate observations permit proper validation of coupled physical-biochemical models and, in turn, identification of their strengths and weaknesses, leading to further development and refinement. The MONCOZE system is highly suitable both for advancing marine science and for the implementation of marine monitoring systems in the context of the joint EU and ESA initiative on GMES (Hackett *et al.*, 2006).

Advanced observing and modelling systems, such as those integrated and implemented in the MONCOZE project, can supply useful information on the ocean and coastal marine environment, as illustrated in Figure 11.1, which shows its organization into three modules: observations, models and information. The operational system encompasses both near-real-time and delayed-mode data, which are bound together by efficient real-time data connections. Regular validation of the system is necessary in order to pin down model deficiencies and, in turn, support the development of the models with the aim of improving their predictive abilities. Data assimilation is also required to produce an optimal four-dimensional description of the physical state and ecosystem and provide initial conditions for forecasts. Finally, the large amount of data produced by the observing and modelling systems requires a dedicated analysis and dissemination facility to provide tailored information to users.

A major challenge facing the efficient operation of such systems is to ensure that the information is available for the broad range of users at their diverse temporal needs. Meeting this challenge requires interaction and feedback among data providers, service providers and end-users. What is more, providing information within the Norwegian exclusive economic area will rely to an ever-increasing extent on national and international co-operation. Data providers include the operators of marine observation systems (e.g. the tide gauge network, monitoring satellites) and the operators of ocean modelling systems. There are several such providers in Norway, some of whom are delivering data in near-real-time. However, non-national sources are important, especially for satellite data and for global model data. Service providers take data products and integrate them into new, value-added products for targeted end-users. The purpose of the service provider is to transform data into information, using analytical tools and human expertise to tailor the information products to specific uses. Scientific research, fisheries, aquaculture, shipping, offshore industry, environmental agencies, coastal management authorities, the Coast Guard, the Navy and recreation are important end-user groups in Norway.

11.3 Possibilities and outlook

It is in the research-driven development and interplay of data provision, service provision and information use that operational oceanography is currently evolving, with the goal of advancing the ability to provide reliable hindcasts (re-analyses), nowcast analyses and forecasts. In this complex process, information technology plays an important role, both in data production (e.g. numerical models) and as a vehicle for accessing and integrating data and information products.

We may take the recent MONCOZE project (implemented around the concept shown in Figure 11.1) as an illustrative example of the current status of opera-

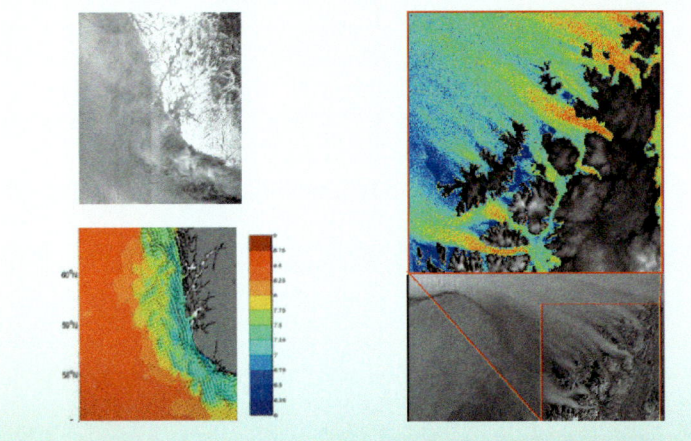

► **Figure 11.2**
(top left) SAR Radar image of the Norwegian Coastal Current on 9 May 2003.
(lower left) Corresponding simulated surface current and sea surface temperature using a numerical model.
(right) High resolution wind field pattern with rapid shifts of 15 m/s derived from radar image off Northern Norway.

tional oceanography for the Norwegian coastal zone, its complexity, strengths and weaknesses, and outlook for the future. A number of satellite-derived data products are included in the Observation module. One is sea surface temperature (SST) maps. These maps are produced twice daily at Meteorological Institute using raw data provided by the United States National Oceanic and Atmospheric Administration (NOAA). The products are developed and maintained under the auspices of the European Organisation for the Exploitation of Meteorological Satellites (EUMETSAT). A second example is the cloud- and light-independent Synthetic Aperture Radar (SAR) images supplied by ESA and processed and analysed by NERSC to derive information about surface current features and high-resolution wind fields (see Figure 11.2) using recently developed methods (Johannessen *et al.*, 2005).

The SAR image expression (Figure 11.2, top left) is fairly complex, with bands of bright and dark radar intensity values mostly confined to the area occupied by the Norwegian Coastal Current. The dominant wavelength of the modulation pattern is about 20–30 km. In comparison, the mesoscale variability of the surface current also has a dominant length scale of 30 km (Figure 11.2, lower left). The current strength associated with these features ranges from 0.4 to 0.6 m/s, and includes an abundance of zones of current convergence and divergence that seem to be manifested in the SAR image. The SAR image (Figure 11.2, right) off the coastal region off northern Norway, on the other hand, shows distinct northwestward-elongated offshore bands of brighter-darker radar signals. Assuming a wind direction from the southeast, the radar image is converted to a wind-speed map (large inset in colour) that clearly shows rapid and intense wind shifts connected with the distinct bands in the radar signals. Wind speeds differ across these bands by up to 15 m/s over distances of less than 500 m.

Near-real-time monitoring of algal bloom and phytoplankton distribution is routinely performed by NERSC using ocean colour satellite data (such as from SeaWiFS). Under cloud-free conditions they are an efficient tool for monitoring the marine ecosystem, as illustrated in Figure 11.3.

The left image shows how a combination of the spectral channels of the SeaWiFS sensor provides a virtually visual image of the ocean surface, resolving large colour contrasts. This confirms that the oceans contain colour-producing material with an extensive spatial and temporal variability. Some parts of the oceans in blue mark clearer water, while various algae blooms can turn the water almost milky white (coccoliths) or nearly red (some flagellates). These colouring agents are usually associated with an abundance of phytoplankton, sediments or dissolved organic matter. Based on the fact that these colour-producing agents have different spectral characteristics, algorithms combining the spectral intensity of the satellite data are applied to estimate phytoplankton concentrations, (Figure 11.3, right), suspended matter and dissolved organic compounds in the water under cloud-free conditions.

In-situ observations include, for example, temperature and salinity profile data from research vessels, some of which are transmitted in near-real-time, profiling floats and high-resolution surface current data from coastal high-frequency (HF) radar systems (Figure 11.4).

Shore-based HF radars provide wide-area coverage of surface currents, and are thus capable of filling a gap that is not satisfactorily covered by satellite remote sensing and *in-situ* observations. The radial component of the surface current is inferred from the Doppler shift observations, and the radial currents derived from two different overlapping radars may be combined to yield the total current vectors. The range and resolution of HF radars depend on the carrier frequency. Typical figures are 50 km radial range and 2 km resolution, although the range can be extended to as much as 200 km by trade-off against lower resolution. HF radars are ideal for coastal ocean observation systems providing regular sampling and extensive spatial coverage, and returning data in real-time (typically a few minutes after acquisition).

The hindcast-analysis-forecast module of Figure 11.1 consists of numerical models run by three Norwegian institutes. For example, operational models

▶ **Figure 11.3**
Ocean colour (SeaWiFS) image covering the North Sea, Skagerrak and coastal waters of southern Norway obtained on 16 July 2003.
(left): Combination of channels (wave lengths).
(right): Chlorophyll *a* concentration and distribution, white marks clouds.

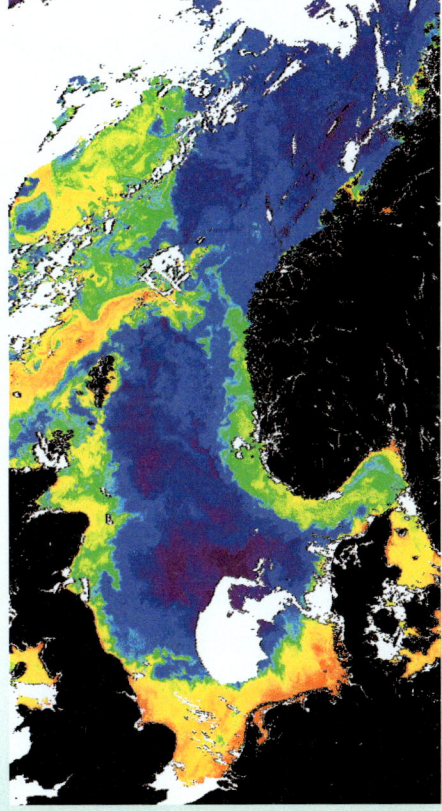

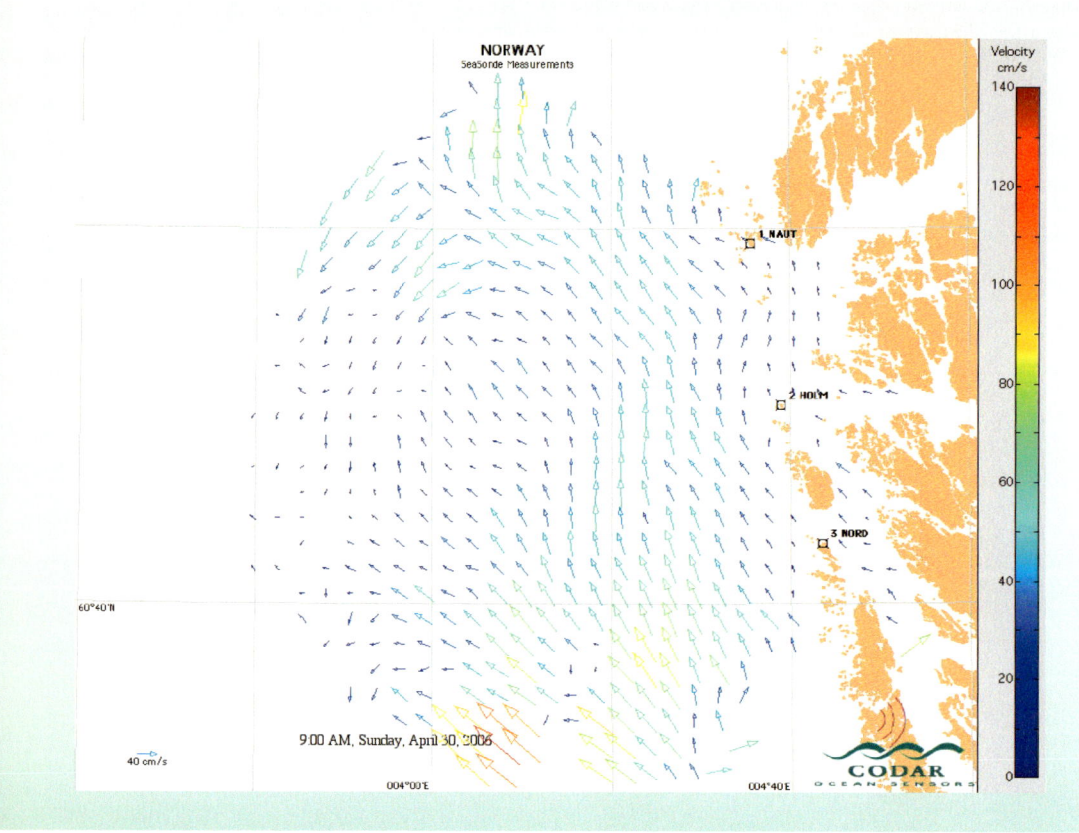

▶ Figure 11.4
Surface current vector map west of Fedje derived from HF radar on 30 April 2006. In the south the current exceeds 1 m/s and further north a clear indication of a cyclonic eddy is seen with a diameter of about 30 km.

for hydrodynamics, nutrients and algae are run for daily forecasting, using atmospheric forcing data from the European Centre for Medium-range Weather Forecasting (ECMWF). For assessments of the state of the ecosystem for management purposes, IMR performs annual hindcast simulations (re-calculation) using the best available forcing data for the previous year.

As a first step in integrating these diverse data types into the Information module, the Pilot Ocean Monitoring System (POMS) was built to accommodate operational products (e.g. satellite SST, model forecasts) and off-line products (e.g. annual ecosystem assessment), as well as providing a limited archival facility (Figure 11.5). One essential design requirement of POMS is the ability to view and compare different products easily. Another is the analysis module, which provides an expert analysis of the current situation, rather like a weather forecaster. A simple web interface makes the information accessible to a wide range of users. By using a common graphics engine for all data and rendering via a Web Map Server (WMS) mapping functions and overlaying can be made available on any web browser (Figure 11.6).

The MONCOZE project has demonstrated that there are useful data products available from ocean model and observing systems, and that even simple means of presenting them together can give value-added information to users. However, there is a need to further develop the integration of data products into information products and to enhance capabilities for information provision. In the near future, we can foresee a further development in the direction of merged and targeted data products using data fusion techniques. Of particular interest is instrumentation for biochemical parameters.

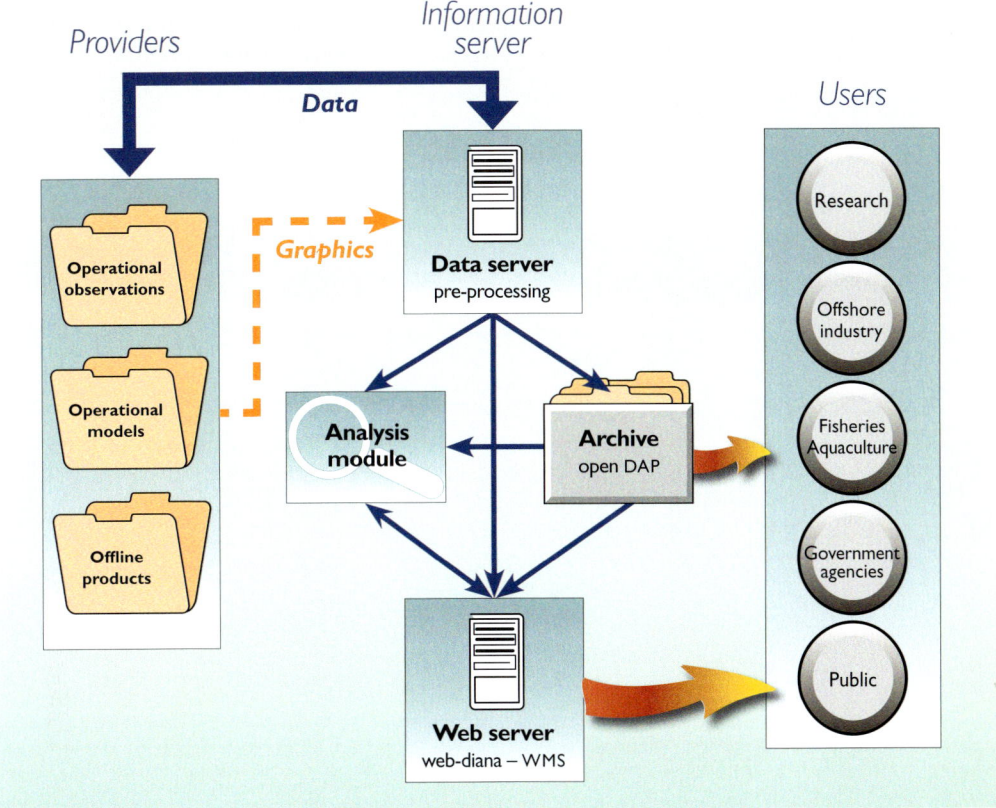

▶ Figure 11.5
Schematic of the MONCOZE Pilot Ocean Monitoring System (POMS).

Graphics: Harald E. Tørresen

Moreover, HF radar technology has matured to the stage where, for example, it is feasible to observe the surface currents across the Skagerrak at two-hourly intervals at ~10 km resolution. Operations of new autonomous floats and gliders for subsurface mapping of the physical and biogeochemical quantities are now providing routine products for operational oceanography. However, the investment that would be required to purchase, operate and sustain such standard and new observation technology for the entire Norwegian exclusive economic zone and coastal areas is a major challenge. It is therefore only feasible to establish, operate and incrementally expand observatories in selected local tie-point regions. A common additional requirement for all observations is ready access in near-real-time.

Improved observation system capabilities will also improve models and model predictions. Modelling capabilities are continually improving, due to expanding computing capacity, which permits better spatial resolution and coverage. However, the models require observations for validation and testing of new techniques, as well as for assimilation of observations in order to improve the forecasts. In the short term (two to three years), operational models will provide forecasts of hydrodynamics, sea-ice and ecosystem parameters up to ten days ahead for the Norwegian exclusive economic zone, at a resolution of 4–10 km. For selected fjord and coastal areas, even finer-scale models will be used (grid-spacing of ~100 m). These local models will obtain operational data on the open boundaries from regional and global models run either nationally or by international partners. International collaboration on global data is thus essential in a cost-benefit perspective.

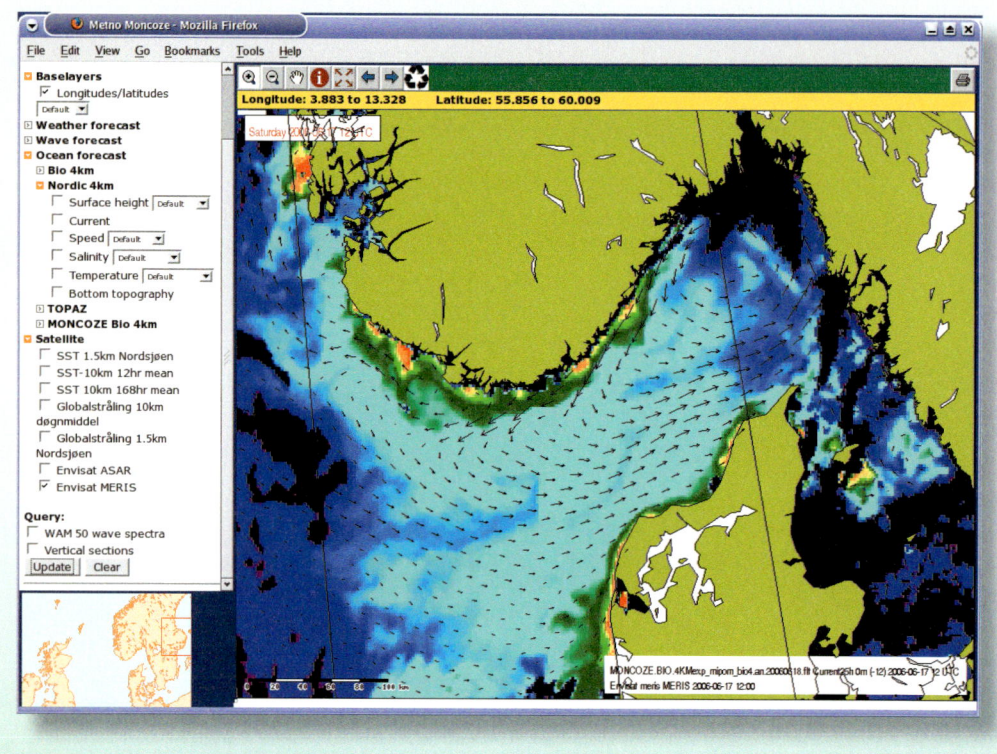

▶ Figure 11.6
Screen shot of prototype POMS Web Map Server (WMS) web page showing a chlorophyll *a* distribution overlaying surface current vectors. Common WMS capabilities such as zooming, panning, point information, etc., are actuated by buttons on the upper toolbar. Available data and background layers are listed in the menu at left, which is regularly updated as new data arrive in the data server.

Information systems such as POMS will need increased capacity and facilities if they are to keep abreast of the upstream developments in data provision and modelling and the downstream requirements of users. Clearly, an information system must be able to access as much observational data as possible, both real-time and archived data. This is a challenging issue that is partly technical and partly organizational, but which also fundamentally relies on commitments to maintaining continuity in data supply. The physical and ecological processes are often non-linear, and assimilation of observations (Evensen, 2003; Bertino *et al.*, 2003) is therefore essential. In view of the increases in computing power new perspectives are emerging for the application of such advanced assimilation techniques in high-resolution, non-linear, models.

The information system must also be able to access a variety of numerical model data-sets, including a range of operational forecasts from both national and other service providers. Access to ancillary data, such as weather forecasts and projected satellite coverage maps, is necessary for a full view of the state of the marine environment. Furthermore, the inclusion of background information, such as climatological data, re-analysis data, bathymetry, seabed characteristics, etc, is required to assess the present state in the context of climate and human activities.

The Web Map Server (WMS) approach (Figure 11.6) to disseminating information products in a flexible manner is proving to be efficient among service providers in many fields. Through compliance with international standards it has the additional advantage of facilitating the exchange of graphical renderings – not the data – between collaborating servers, thereby opening new avenues for special user-defined, integrated information products. New technologies will enable users to find relevant information, view a variety of information products with considerable flexibility in form and fetch data sets using a web browser. They will also allow service providers to tailor information servers to the needs of specific user groups.

Complementing the expected advances in operational oceanography systems in the coming decade, there is a significant need to create a new curriculum at university level to meet the growing demand for recruitment in this field. Presently, courses in operational oceanography are offered at the Universities of Bergen and Oslo, and there are considerations for a subsequent Masters degree program in collaboration and partnership with foreign institutes and universities. A plan for a dedicated Research School in Operational Oceanography has also emerged. Provided priorities are agreed and sufficient funding is allocated, it is expected that within three to five years Norway can provide a leading initiative in the development and implementation of such a curriculum in operational oceanography.

Finally, operational oceanography is now moving into an era in which it will be opportune to exploit its potential for the benefit of society. On a European level, the joint EU and ESA Global Monitoring System for Environment and Security (GMES) initiative puts the need for establishment and development of integrated observing and modelling systems at the forefront, primarily with a focus on operational services rather than research. At present, the EU is formulating a new European Marine Strategy. A major issue to be tackled within this strategy is the need to define better ways to extract, in a sustainable manner, greater benefit to European citizens from the oceans. In so doing, operational oceanography is highly relevant in, for instance, the context of:

- ecosystem-based management of the marine environment for coastal waters on a regional basis
- scientific research on marine biology, energy, maritime transportation, fisheries, environment and aquaculture
- the provision of reliable forecasts for tomorrow's weather, forthcoming seasons, the direction and speed of climate change, and extreme events related to flooding and subsequent heavy river runoff, storm surges, hazardous winds, waves and rip currents, and harmful algal blooms.

Operational oceanography thus plays an essential role in the interleaved cycles between marine policy formulation, implementation, review and updating on the one hand and environmental monitoring, rapid assessment, scientific analyses, reporting and decision support on the other. In particular, quantitative information in the form of tailored indicators will be regularly provided by operational oceanography systems to document and explain how environmental quality undergoes temporal and spatial variations. Over time, such indicators will in turn be a vital factor in focusing on and illuminating the significance of environmental change, and in defining and implementing adequate short-term actions that, in the longer term, can ensure that progress is made in the direction of sustainable management of the marine environment.

Relevant servers and web-sites
- *http://hab.nersc.no* – near-real-time monitoring of algal blooms and phytoplankton distribution, carried out by NERSC.
- *http://hf.met.no* – near-real-time high-resolution surface current vector map west of Fedje on the coast of Western Norway, derived from CODAR system

and displayed and disseminated by the Norwegian Meteorological Institute.
- *http://met.no/kyst_og_hav/havvarsel.html* – daily updated forecasts of currents, temperature and salinity provided by the Norwegian Meteorological Institute.
- *http://topaz.nersc.no/* – weekly analyses and seven-day forecasts of currents, temperature, salinity, sea ice extent, etc. provided by the Mohn-Sverdrup Center/NERSC.
- *http://www.coriolis.eu.org* – downloads of operational Argo profiling float data.
- *http://www.eurogoos.org* – information on the European contribution to GOOS.
- *http://www.geo.org/* – information on GEOSS.
- *http://www.gmes.info* – information on GMES.
- *http://www.mersea.eu.org* – information on the MERSEA Integrated Project and with data display/download capabilities.
- *http://moncoze.met.no* – visualization of operational MONCOZE products.
- *http://www.nersc.no/MONCOZE* – information on the MONCOZE project.
- *http://www.noos.cc/* – information on NOOS.
- *http://www.usgodae.org/* – information on GODAE. Figure legends.

References

Anon 1886. Ergebnisse der Untersuchungsfahrten "Drache" in der Nordsee in den Sommern 1881, 1882 und 1884. Veröffentlichungen des Hydrographischen Amts der K. u. K. Kriegsmarine in Pola, Berlin.

Anon 1970. Joint Skagerrak Expedition 1966, Vol. 1–5. Conseil International pour l'Exploration de la Mer, Service Hydrographique, Andr. Fred. Høst & Son, Copenhagen.

Anon 1975. Litteraturliste for norske kystfarvann. Fysisk, kjemisk og biologisk oseanografi samt marin geologi. Rapport nr. 5 fra Norsk Oseanografisk Datasenter. April 1975: 26 pp.

Anon 1976. Some preliminary results from a synoptic experiment in the Norwegian Coastal Current (SEX 75). Report 1/76. The Norwegian Coastal Current Project, 34 pp, 73 figs.

Anon 1979. Remote Sensing Experiment in the Norwegian Coastal Waters. Spring 1979. Report 3/79. The Norwegian Coastal Current Project, 26 pp, 38 figs.

Anon 1993. North Sea – Subregion 8. Assessment report 1993. North Sea Task Force. State Pollution Control Authority (SFT), Norway.

Anon 1997. Den norske Los I. Alminnelige opplysninger. Statens Kartverk Sjøkartverket, 240 pp.

Asplin, L., Salvanes, A.G.V. and Kristoffersen, J.B. 1999. Non-local wind-driven fjord-coast advection and its potential effect on plankton and fish recruitment, Fisheries Oceanography, 8: 255–263.

Asvall, R.P. 1976. Effects of regulations on freshwater runoff. Pp. 15–20 in: Skreslett, S., Leinebø, R., Matthews, J.B.L. and Sakshaug, E. (eds), Fresh water on the sea. The Association of Norwegian Oceanographers, Oslo 1976.

Audunson, T., Dalen, V., Krogstad, H., Lie, H.N. and Steinbakke, P. 1981. Some observations of ocean fronts, waves and currents in the surface along the Norwegian coast from satellite images and drifting buoys. Pp. 20–56 in: The Norwegian Coastal Current, R. Sætre and M. Mork (eds), University of Bergen.

Aure, J. and Sætre, R. 1981. Wind effects on the Skagerrak outflow. Pp. 263–293 in: The Norwegian Coastal Current, R. Sætre and M. Mork (eds). University of Bergen.

Aure, J., Svendsen, E., Rey, F. and Skjoldal, H.R. 1990. The Jutland Current: Nutrient and physical oceanographic conditions in late autumn 1989. ICES CM 1990/C: 35, 12 pp.

Aure, J. and Østensen, Ø. 1993. Hydrographic normals and long-term variations in Norwegian Coastal waters. Fisken og Havet, 6: 75 pp.

Aure, J., Molvær, J. and Stigebrandt, A. 1996. Observations of inshore water exchange forced by a fluctuating offshore density field. Marine Pollution Bulletin, 33(1–6): 112–119.

Aure, J., Danielssen, D. and Sætre, R. 1996. Assessment of eutrophication in Skagerrak coastal waters using oxygen consumptions in fjordic basins. ICES Journal of Marine Science, 53: 589–595.

Aure, J., Danielssen, D. and Svendsen, E. 1998. The origin of Skagerrak coastal water off Arendal in relation to variations in nutrient concentrations. ICES Journal of Marine Science, 55: 610–619.

Aure, J., Strand, Ø. and Strohmeier, T. 2005. Improved production of mussels by artificial upwelling of nutrient rich deep water. Report Institute of Marine Research, October 2005, 12 pp. (In Norwegian).

Bang-Andersen, S. 2003: Southwest Norway at the Pleistocene/Holocene Transition: Landscape Development, Colonization, Site Types, Settlement Patterns. Norwegian Archaeological Review, 36, 1/2003, 5–25.

Bertino, L. and Evensen, G. 2003. The DIADEM/TOPAZ monitoring and prediction system for the North Atlantic. In: Proceedings of the Third International Conference on EuroGOOS, 3–6 December 2002, Athens, Greece, ed. by H. Dahlin, N.C. Flemming, K. Nittis, S.E. Petersson, Elsevier Oceanography Series, 69.

Bertino, L., Evensen, G. and Wackernagel. H. 2003. Sequential data assimilation techniques in oceanography. International Statistical Review. 71 (2): pp. 223–241.

Bjørnsen, V. 2003. Naturvitenskap og politikk. Den norske Nordhavsekspedisjonen 1876–78. Hovedoppgave i historie. Institutt for historie, Universitetet i Tromsø. 120 pp.

Bleck, R. 2002. An oceanic general circulation model framed in hybrid isopycnic-Cartesian coordinates. Ocean Modelling, 4: 55-88.

Blindheim, J., Loeng, H. and Sætre, R. 1981. Long-term temperature trends in Norwegian Coastal waters. ICES, C.M. 1981/C: 19, 13 pp.

Blindheim, J. 2004. Oceanography and climate. Pp. 65–96 in: The Norwegian Sea Ecosystem. Ed. by H.R. Skjoldal, R. Sætre, A. Fernö, O.A. Misund and I. Røttingen. Tapir Academic Press, Trondheim.

Blumberg, A. and Mellor, G. 1987. A description of the three-dimensional coastal ocean circulation model. In: Three-dimensional Coastal Ocean Models, ed. by N. Heaps, vol. 4 of Coastal and Estuarine Sciences, American Geophysical Union.

Bondevik, S., Løvholt, F., Harbitz, C., Mangerud, J., Dawson, A. and Svendsen, J.I. 2005. The Storegga Slide tsunami – comparing field observations with numerical simulations. Marine and Petroleum Geology, 22: 195–2008.

Buhl-Mortensen, L., Aure, J., Alve, E., Husum, K. and Oug, E. 2006. Effects of hypoxia on fjordfauna: The bottom fauna and environment of fjords on the Skagerrak coast. Fisken og Havet, nr. 3, 108 pp. Institute of Marine Research. (In Norwegian).

Dalpadado, P. 1989. Bibliografi over litteratur om kystøkologi. Report BKO 8905. Institute of Marine Research, Bergen.

Danielssen, D.S., Svendsen, E. and Ostrowski, M. 1996. Long-term hydrographic variations in the Skagerrak based on the section Torungen–Hirtshals. ICES Journal of Marine Science, 53: 917–925.

Dannevig, P. 1992. Vær, vind og sjø på norskekysten. Nordanger Forlag, 4. utgave, 64 pp.

Dickson, R.R., Meincke, J., Malmberg, S. Aa. and Lee, A. 1988. The "Great Salinity Anomaly" in the Northern North Atlantic 1968–1982. Progressive Oceanography, 20: 103–151.

Dickson, R.R. and Meincke, J. 2003. The North Atlantic Oscillation and the ocean's response in the 1990s. ICES Marine Science Symposia, 219: 15–24.

Dommasnes, A., Rey, F. and Røttingen, I. 1994. Reduced oxygen concentration in herring wintering areas. ICES Journal of Marine Science, 51: 63–69.

Dundas, I., Johannessen, O.M., Berge, G. and Heimdal, B. 1989. Toxic Algal Bloom in Scandinavian Waters, May–June 1988. Oceanography. Off. mag. of Oceanographic Society, Vol. 2, No. 1.

Dybern, B.I., Danielssen, D.S., Hernroth, L. and Svendsen, E. 1994. The Skagerrak Experiment – Skagex Report 1988–1994. Nordic Council of Ministers, Copenhagen. ISSN 92-9120-565.

Eggvin, J. 1932. Vannlagene på fiskefeltet. Årsberetning vedkommende Norges fiskerier, 2: 90–95.

Eggvin, J. 1936. Oceanografiske forhold i Nordnorge under lofot- og finnmarksfisket 1936. En foreløbig utredning. Årsberetning vedkommende Norges fiskerier, 2: 131–143.

Eggvin, J. 1938. Trekk fra Nord-Norges oceanografi sett i sammenheng med torskefisket. Fiskeridirektoratets Skrifter, Serie Havundersøkelser, 5(7): 33–46.

Eggvin, J. 1940. The movement of a cold water front. Fiskeridirektoratets Skrifter, Serie Havundersøkelser, 6(5): 151 pp.

Eggvin, J. 1943. The great exchange of water masses along the Norwegian coast, 1940. Rapports et Procès-Verbaux des Réunions, 112: 49–61.

Eggvin, J. 1946. Den fysiske oseanografi og våre fiskerier. Forskning og Fremsteg: 69–89, Bergen.

Eide, E.A. 2002. Introduction – Plate reconstruction and integrated datasets. In: Eide, E.A. (coord.), BATLAS – Mid-Norway plate reconstruction atlas with global and Atlantic perspectives. Geological Survey of Norway, pp. 8–17.

Ekman, G. 1870. Om hafsvattnet utmed Bohuslänska kusten. Kungliga Svenska Vetenskaps-Akademien Handlingar, 9(4): 44 pp.

Engedahl, H. 1995. Implementation of the Princeton Ocean Model (POM/ECOM3D) at the Norwegian Meteorological Institute. Research Report No. 5, Norwegian Meteorological Institute, Oslo, Norway.

Engedahl, H., Lunde, A., Melsom A. and Shi, X.B. 2001. New schemes for vertical mixing in MI-POM and MICOM. Research Report No. 118, Norwegian Meteorological Institute, Oslo, Norway. 51 pp.

Evensen, G. 2003. The ensemble Kalman filter: theoretical formulation and practical implementation, Ocean Dynamics, 53: 343–367.

Førland, E.J., Roald, L.A., Tveito, O.E. and Hanssen-Bauer, I. 2000. Past and future variations in climate and runoff in Norway. Norwegian Meteorological Institute. Report No. 19/00 KLIMA, 77 pp.

Førland E.J. and Nordeng, T.E. 1999. Framtidig klimautvikling i Norge. Cicero, 6/99: 21–24.

Førland, E., Roald, L.A., Tveito, O.E. and Hanssen-Bauer, I. 2000. Past and future variations in climate and runoff in Norway. Klima, Report no. 19, 77 pp.

Føyn, L. and Rey, F. 1981. Nutrient distribution along the Norwegian Coastal Current. Pp. 629–639 in: The Norwegian Coastal Current. R. Sætre and M. Mork (eds). University of Bergen, 1981.

Gade, G. 1894. Temperaturmålinger i Lofoten 1891–1892. Carl C. Werner & Co, Christiania, 162 pp.

Gade, H.G. 1986. Features of fjord and ocean interaction. Pp. 182–189 in: The Nordic Seas. Ed. B.G. Hurdle, Springer Verlag, New York.

Gammelsrød, T. and Sætre Hjøllo, S. 2005. Stabiliteten i Golfstrømsystemet. Cicero 5/2005: 20–22.

Gjevik, B., Moe, H.D. and Ommundsen, A. 1997. Sources of the Maelstrom. Nature, 388: 837–838.

Gran, H.H. 1900. Hydrobiological studies of the Northern Ocean and the coast of Nordland. Report on Norwegian Fishery and Marine Investigations. 1(5): 92 pp.

Gustafsson, B. 1997. Dynamics of the sea and straits between the Baltic and North Sea – A process-oriented oceanographic study. Thesis, Department of Oceanography, University of Gothenburg.

Hackett, B., Røed, L.P., Gjevik, B., Martinsen, E.A. and Eide, L.I. 1995. A Review of the Metocean Modelling Project (MOMOP). Part 2: Model

Validation Study. In: Quantitative Skill Assessment for Coastal Ocean Models, ed. by D.R. Lynch and A.M. Davies, Coastal and Estuarine Studies Vol. 47, American Geophysical Union, 307–327.

Hackett, B., Albretsen, J., Røed, L.P., Johannessen J.A. and Svendsen, E. 2006. The MONCOZE Pilot Ocean Monitoring System (POMS); A Tool for Marine Environmental Monitoring. European Operational Oceanography: Present and future. In: Proceedings of 4th International Conference on EuroGOOS. Ed. by H. Dahlin, N.C. Flemming, P. Marchand and S.E. Petersson, European Communities, 242–247.

Hátun, H., Sandø, A.B., Drange, H., Hansen, B. and Valdemarsen, H. 2005. Influence of the Atlantic Subpolar Gyre on the thermohaline circulation. Science, 309: 1841–1844.

Heyen, H., Zorita, E. and von Storch, H. 1996. Statistical downscaling of monthly mean North Atlantic air-pressure to sea level anomalies in the Baltic Sea. Tellus, 48A: 312–323.

Helland-Hansen, B. and Nansen, F. 1909. The Norwegian Sea. Its physical oceanography based upon the Norwegian researches 1900–1904. Report on Norwegian Fishery and Marine Investigations. 2(2): 390 pp.

Helland-Hansen, B. 1930. Havet og havstraumane. "Den 7de juni", Stord. 30 pp.

Hjort, J. and Gran, H.H. 1899. Currents and pelagic life in the Northern Ocean. Report on Norwegian Fishery and Marine Investigations 1895–1897. Bergen Museums Skrifter Nr. 6: 22 pp.

Hjort, J. and Gran, H.H. 1900. Hydrographic-biological investigations of the Skagerrak and the Christiania fiord. Report on Norwegian Marine Investigations, 1(2): 56 pp.

Hognestad, P.T. 1973. Forsøk med strømflasker i Nord-Norge i 1972. Fiskets Gang, 14: 289–293.

Holtedahl, H. 1993. Marine geology of the Norwegian continental margin. Norges geologiske undersøkelse, Special publication No. 6, 150 pp.

Hurrel, J.W. 1995. Decadal trends in the North Atlantic Oscillation: Regional temperatures and precipitation. Science, 269: 676–679.

Iden, K.A. 1997. Meteorologi – Værforholdene på norskekysten. Pp. 129–146 in: Den norske Los, Alminnelige opplysninger 1:240 pp.

Ingvaldsen, R., Loeng, H., Ådlandsvik, B. and Ottersen, G. 2003. Climate variability in the Barents Sea during the 20th century with a focus on the 1990s. ICES Marine Science Symposia, 219:160–168.

James, I.D. and McClimans, T.A. 1983. Coastal current whirls in laboratory and numerical models, Ocean Modelling, 53: 1–3.

Johannesen, O.M., Johannessen, J.A., Sandven, S. and Davidson, K.L. 1986. Preliminary results of the Marginal Ice Zone Experiment (MIZEX) summer operations. Pp. 665–679 in: The Nordic Seas, ed. by B.G. Hurdle, Springer Verlag, New York.

Johannessen, O.M. and Pettersson, L.H. 1988. Report from a meeting on an operational Ocean Monitoring and Forecasting system (HOV). (In Norwegian). Conference Report no. 1, February 1988, Nansen Environmental and Remote Sensing Center.

Johannessen, J.A., Kudryavtsev, V., Akimov, D., Eldevik, T., Winther, N. and Chapron, B. 2005. On Radar Imaging of Current Features; 2: Mesoscale Eddy and Current Front detection. Journal of Geophysical Research, 110: C07017, doi: 10.1029/2004JC002802.

Johannessen, J.A., Hackett, B., Svendsen, E., Søiland, H., Røed, L.P., Winther, N., Albretsen, J., Danielssen, D., Pettersson, L., Skogen, M. and Bertino, L. 2006. Monitoring the Norwegian Coastal Zone Environment – The MONCOZE Approach, European Operational Oceanography: Present and future. Pp. 809–815 in: Proceedings of 4th International Conference on EuroGOOS. Ed. by H. Dahlin, N.C. Flemming, P. Marchand and S.E. Petersson, European Communities.

Johannessen, J.A., Svendsen, E., Sandven, S., Johannessen O.M. and Lygre, K. 1989. Three Dimensional Structure of Mesoscale Eddies in the Norwegian Coastal Current. Journal of Physical Oceanography, 19 (1): 3–19.

Johannessen, J.A., Svendsen, E., Sandven, S., Johannesen O.M. and Lygre, K. 1989. Three-dimensional structure of mesoscale eddies in the Norwegian Coastal Current. Journal of Physical Oceanography, 19: 3–19.

Johannessen, J.A., Shuchman, R.A., Digranes, G., Lyzenga, D.R., Wackerman, C., Johannessen, O.M. and Vachon, P.W. 1996. Coastal ocean fronts and eddies imaged with ERS-1 synthetic aperture radar. Journal of Geophysical Research, 110(C3): 6651–6667.

Kahma, K.K., Boman, H., Johansson, M.M. and Launiainen, J. 2003. The North Atlantic Oscillation and sea level variations in the Baltic Sea. ICES Marine Science Symposia, 219: 365–366.

Knudsen, M. 1899. De hydrografiske forhold i de danske farvande indenfor Skagen i 1894–1898. Beretning fra Kommisjonen for videnskabelige undersøgelser av de danske farvande. Copenhagen, 2: 19–79. (In Danish).

Longva, O. and Thorsnes, T. (eds). 1997. Skagerrak in the past and at the present – an integrated study of geology, chemistry, hydrography and microfossil ecology. Norges geologiske undersøkelse, Special Publication 8, 100 pp.

Ljøen, R. and Sætre, R. 1978. Long-term hydrographic variations off southern Norway. Rapports et Procès-Verbaux des Réunions du Conseil International pour l'Exploration de la Mer, 172: 345–349.

Ljøen, R. 1970. Kalde vintres innflytelse på de hydrografiske forhold i Nordsjø–Skagerrak-området. Fiskets Gang, 21: 394–400.

Ljøen, R. 1980. Atlas of mean temperature, salinity and density in the summer from the northern North Sea. Fisken og Havet, 2: 37 pp.

Ljøen, R. 1981. Seasonal variations in the inflow of different water masses to the Skagerrak. Pp. 357–369 in: The Norwegian Coastal Current. Sætre, R. and Mork, M. (eds). University of Bergen.

Maunula, P. (2006a). Annual variation of nitrate concentrations in the Arkona Basin. In: The Baltic Sea Portal. www.balticseaportal.fi.

Maunula, P. (2006b). Annual variation of phytoplankton in the Arkona Basin. In: The Baltic Sea Portal. www.balticseaportal.fi.

Martinsen, E., Hackett, B., Røed L.P. and Melsom, A. 1997. Operational marine models at the Norwegian Meteorological Institute. In: Operational Oceanography, The challenge for European Co-operation. Ed. by J.H. Stel, H.W.A. Behrens, J.C. Borst, L.C. Droppert, J.P. van der Meulen, Elsevier Oceanography Series, 62, The Netherlands.

McClimans, T.A. and Johannessen, B.O. 1998. On the use of laboratory ocean circulation models to simulate mesoscale (10–100 km) spreading. Environmental Modelling & Software, 13: 443–453.

Mitchelson-Jacob, G. and Sundby, S. 2001. Eddies of Vestfjorden, Norway. Continental Shelf Research, 21: 1901–1918.

Midttun, L. 1971. Long-term observation series of surface temperature and salinity in Norwegian waters. ICES CM 1971/ C:25, 3 pp.

Mohn, H. 1887. The North Ocean, its depths, temperature and circulation. In: The Norwegian North-Atlantic Expedition 1876–1878. Vol. 2. Grøndahl & Søn, Christiania. 212 pp.

Mork, M. 1989. Oceanography in Norway: fragments of its history over the last 100 years. Norsk Geografisk Tidsskrift, 43: 129–133.

Mysak, L.A. and Schott, F. 1977. Evidence of baroclinic instability of the Norwegian Current. Journal of Geophysical Research, 82(15): 2087–2095.

Ottesen, D., Dowdeswell, J.A. and Rise, L. 2005. Submarine landforms and the reconstruction of fast-flowing ice streams within a large Quaternary ice sheet: The 2500 km-long Norwegian–Svalbard margin (57–80°N). Geological Society of America Bulletin, 117: 1033–1050.

Orvik, K.A., Skagseth, Ø. and Mork, M. 2001. Atlantic inflow to the Nordic Seas: current structure and volume fluxes from moored current meters, VM-ADCP and SeaSoar-CTD observations 1995–1999. Deep-Sea Research I, 48: 937–957.

Orvik, K.A. and Skagseth, Ø. 2005. Golfstrømmen er blitt varmere og svakere de 10 siste år. Cicero 5/2005: 18–19. (In Norwegian).

Pettersson, O. and Ekman, G. 1891. Grunddragen af Skageracks och Kattegats hydrografi. Kungliga Svenska Vetenskaps-Akademien Handlingar, 24(11): 161 pp. (In Swedish).

Pettersson, O. and Ekman, G. 1897. De hydrografiske förändringarne innom Nordsjöns och Östersjöns område. Kungliga Svenska Vetenskaps-Akademien Handlingar, 29(5): 125 pp. (In Swedish).

Pontoppidan, E. 1755. The Natural History of Norway, London 1755. 291 pp.

Prøsch-Danielsen, L. and Høgestøl, M. 1995: A coastal Ahrensburgian site found at Galta, Rennesøy, Southwest Norway. Pp. 132–130 in: A. Fisher (ed.) Man and sea in the Mesolithic. Oxbow Monograph 53, Oxbow Books, Oxford.

Prøsch-Danielsen, L., Høgestøl, M. and Bøe, R. 2005. Undersjøiske skred og flodbølger (tsunamier) i Boknafjorden – da steinalderlokaliteten på Galta ble skylt på havet. Frá haug ok heidni 1/2005, 23–29.

Rinde, E., Bjørge, A., Eggereide, A. and Tufteland, G., (eds) 1998. Kystøkologi – den ressursrike norskekysten. Universitetsforlaget, Oslo, 214 pp.

Rokoengen, K. and Johansen, A.B. 1996. Possibilities for early settlement on the Norwegian Continental shelf. Norsk Geologisk Tidsskrift, 76: 121–125.

Rydberg, L. 1996. Ch. 4.5.2 Hydrography. In: Third periodic assessment of the state of the marine environment of the Baltic Sea, 1989–1993. Baltic Sea Environment Proceedings No. 64B.

Røed, L.P. and Fossum, I. 2004. Mean and eddy motion in the Skagerrak/northern North Sea: insight from a numerical model. Ocean Dynamics, 54(2): 197–220.

Røed, L.P., Hackett, B., Gjevik, B. and Eide, L.I. 1995. A Review of the Metocean Modelling Project (MOMOP). Part 1: Model Comparison Study, pp. 285–305 in: Quantitative Skill Assessment for Coastal Ocean Models. Ed. by D.R. Lynch and A.M. Davies, Coastal and Estuarine Studies Vol. 47, American Geophysical Union.

Sars, G.O. 1879. Indberetninger til Departementet for det Indre fra Professor G.O. Sars om de af ham i årene 1864–1878 anstillede undersøgelser angaaende saltvandsfiskerierne. Berg og Ellefsens Bogtrykkeri, Christiania: 221 pp. (In Norwegian).

Schwach, V. 2000. Havet, fisken og vitenskapen. Fra fiskeriundersøkelser til havforskningsinstitutt 1860–2000. Institute of Marine Research, Bergen: 405 pp. (In Norwegian).

Sherman, K. and Skjoldal, H.R. (eds). 2002. Large marine ecosystems of the North Atlantic. Changing states and sustainability. Elsevier, Amsterdam: 449 pp.

Skjoldal, H.R., Hopkins, C., Erikstad, B. and Leinaas, H.P. (eds). 1995. Ecology of fjords and coastal waters. Proceedings of the Mare Nor Symposium on the Ecology of Fjords and Coastal Waters, Tromsø, Norway, 5–9 December 1994. Elsevier, Amsterdam. 623 pp.

Skarðhamar, J. and Svendsen, H. 2005. Circulation and shelf-ocean interaction off North Norway. Continental Shelf Research, 25: 1541–1560.

Skogen, M.D. 1993. A User's Guide to NORWECOM, The Norwegian Ecological Model System. ISSN 0804–2128. Report no. 6, Centre for Marine Environment, Institute of Marine Research, Bergen.

Skogen, M.D., Svendsen, E., Berntsen, J., Aksnes, D. and Ulvestad, K.B. 1995. Modelling the primary production in the North Sea using a coupled 3-dimensional Physical Chemical Biological Ocean model. Estuarine, Coastal and Shelf Science, 41: 545–565.

Skogen, M.D., Søiland, H. and Svendsen, E. 2004. Effects of changing nutrient loads to the North Sea. Journal of Marine Systems, 46(1–4): 23–38.

Skreslet, S., Leinebø, R., Matthews, J.B.L. and Sakshaug, E. (eds). 1976. Fresh water on the sea. Proceedings from a symposium on the influence of freshwater outflow on biological processes in fjords and coastal waters. 22–25 April 1974, Geilo, Norway. The Association of Norwegian Oceanographers, 246 pp.

Slagstad, D. and Støle-Hansen, K. 1990. Dynamics of plankton growth in the Barents Sea: model studies. Polar Research, 10: 173–186.

Stigebrandt, A. and Aure, J. 1990. De ytre drivkreftenes betydning for vannutskiftningen i fjordene fra Skagerrak til Finnmark. Rapport FO 9003, Senter for marint miljø, Havforskningsinstituttet. (In Norwegian).

Stigebrandt, A. 1990. On the response of the horizontal mean vertical density distribution in a fjord to low-frequency density fluctuations in the coastal water. Tellus, 42A: 605–614.

Sundby, S. 1980. The development of the oceanographic research in Vestfjorden. Fisken og Havet, 1980(1): 11–25.

Sundby, S. 1984. Influence of bottom topography on the circulation at the continental shelf off northern Norway. Fiskeridirektoratets Skrifter, Serie Havundersøkelser, 17: 501–519.

Svendsen, E. 1995. Havstrømkart og biologi i Skagerrak. Internal project report, Project No. 01.06.1, 9 pp.

Svansson, A. 1975. Physical and chemical oceanography of the Skagerrak and the Kattegat. Fishery Board of Sweden, Report no. 1, 1975.

Svansson, A. 1984. Hydrographic features of the Kattegat. Rapports et Procès-Verbaux des Réunions, Conseil Permanent International pour l'Exploration de la Mer, 185: 78–90.

Svendsen, E., Balchen, J.G., Johannessen, J.A., Bogstad, B., Alfredsen, J.A. Slagstad, D., Skogen M. and Tande, K. 2002. The AMOEBE Plan: A model-based and data driven operational ecological biomass estimator, Institute of Marine Research, ISBN: 82-7461-054-7.

Sverdrup, H.U. 1952. Havlære. Vanlig havlære. Vannmasser utenfor norskekysten. Fabritius og Sønners Forlag. 109 pp. (In Norwegian).

Sætre, R. and Ljøen, R. 1972. The Norwegian Coastal Current. Pp. 514–535 in: Stabell Wetteland, S. and Brun, P. (eds). Proceedings of the first international conference on port and ocean engineering under arctic conditions 1. Department of port and ocean engineering, Technical University of Norway, Trondheim.

Sætre, R. 1973. Temperatur og saltholdighetsnormaler for overflatelaget i norske kystfarvann. Fiskets Gang, 8: 166–172. (In Norwegian).

Sætre, R. 1976. Strømflaskeobservasjoner fra Møre–Helgelandsplatået. Report 2/76. The Norwegian Coastal Current Project. 16 pp. (In Norwegian).

Sætre, R. 1978. Midlere temperatur og saltholdighet i overflatelaget langs en del skipsruter i Nordsjøen. Fisken og Havet, Serie B, 1978, nr. 1: 21 pp. (In Norwegian).

Sætre, R. 1979. Features of the mean annual surface salinity variations off southern Norway. Report 1/79. The Norwegian Coastal Current Project, 15 pp.

Sætre, R. 1981. The surface circulation off southern Norway during summer indicated by driftbottles. Report 1/81. The Norwegian Coastal Current Project, 17 pp.

Sætre, R. and Mork, M. (eds) 1981. The Norwegian Coastal Current. Proceedings of the Norwegian Coastal Current Symposium, Geilo, 9–12 September 1980. 795 pp.

Sætre, R. 1983. Strømforholdene i øvre vannlag utenfor Norge. Forskningsprogram om Havforurensning (FOH). Report FO 8306. Institute of Marine Research, Bergen.

Sætre, R., Aure, J. and Ljøen, R. 1988. Wind effects on the lateral extension of the Norwegian Coastal Current. Continental Shelf Research, 8(3): 239–253.

Sætre, R. 1999. Features of the central Norwegian shelf circulation. Continental Shelf Research, 19: 1809–1831.

Sætre, R. and Blindheim, J. 2002. Jens Eggvin – A Norwegian pioneer in operational oceanography. ICES CM 2002/W:01: 20 pp.

Sætre, R., Aure, J. and Danielssen, D.S. 2003. Long-term hydrographic variability patterns off the Norwegian coast and in the Skagerrak. ICES Marine Science Symposia, 219: 150–159.

Sætre, R. 2004. Scientific research in the Norwegian Sea: Background and history, pp. 33–64 in: The Norwegian Sea Ecosystem. Ed. by H.R. Skjoldal, R. Sætre, A. Fernøe, O.A. Misund and I. Røttingen. Tapir Academic Press, Trondheim 2004.

Tollan, A. 1976. River runoff in Norway. Pp. 11–13 in: Skreslett, S., Leinebø, R., Matthews, J.B.L. and Sakshaug, E. (eds). Fresh water on the sea. The Association of Norwegian Oceanographers, Oslo, 1976.

Toresen, R. and Østvedt, O.J. 2000. Variation in abundance of Norwegian spring-spawning herring (*Clupea harengus*, Clupeidae) troughout the 20th century and the influence of climate fluctuations. Fish and fisheries, 1: 231-256.

Torsvik, T.H., Carlos, D., Mosar, M., Cocks, L.R.M. and Malme, T. 2002: Global reconstructions and North Atlantic paleogeography 440 Ma to recent. Pp. 18–39 in: Eide, E.A. (coord.), BATLAS – Mid Norway plate reconstruction atlas with global and Atlantic perspectives. Geological Survey of Norway.

Vikebø, F. 2005. The impact of climate on early stages of Arcto-Norwegian cod – a model approach. Dr.scient.-thesis in geophysics. Geophysical Institute, University of Bergen, 141 pp.

Vorren, T. and Mangerud, J. 2006. Istider kommer og går. In: Ramberg, I.B., Bryhni, I. & Nøttvedt, A. (eds), 2006: Landet blir til – Norges geologi, Norsk Geologisk Forening, Trondheim, pp. xx-xx (In press).

Vorren, T., Mangerud, J., Blikra, L.H., Nesje, A. and Sveian, H. 2006. Norge trer fram. In: Ramberg, I.B., Bryhni, I. & Nøttvedt, A. (eds), 2006: Landet blir til – Norges geologi, Norsk Geologisk Forening, Trondheim, pp. xx-xx (In press).

Wyrtki, K. 1954. Schwankungen im Wasserhaushalt der Ostsee. Deutsche Hydrographische Zeitschrift, 7: 91-129.

Ærtebjerg, G., Andersen, J.H. and Hansen, O.S. (eds). 2003. Nutrients and Eutrophication in Danish Marine Waters. A Challenge for Science and Management. National Environmental Research Institute, Denmark. 126 pp.